Advanced Structured Materials

Volume 23

Series editors

Andreas Öchsner, Southport Queensland, Australia
Lucas F.M. da Silva, Porto, Portugal
Holm Altenbach, Magdeburg, Germany

More information about this series at http://www.springer.com/series/8611

H.D. Mustafa · Sunil H. Karamchandani
Shabbir N. Merchant · Uday B. Desai

tuPOY: Thermally Unstable Partially Oriented Yarns

Silicon of the Future

 Springer

H.D. Mustafa
Department of Electrical Engineering
Indian Institute of Technology Bombay
Mumbai, Maharashtra
India

Shabbir N. Merchant
Department of Electrical Engineering
Indian Institute of Technology Bombay
Mumbai, Maharashtra
India

Sunil H. Karamchandani
Department of Electrical Engineering
Indian Institute of Technology Bombay
Mumbai, Maharashtra
India

Uday B. Desai
Indian Institute of Technology Hyderabad
Hyderabad, Andhra Pradesh
India

ISSN 1869-8433   ISSN 1869-8441   (electronic)
Advanced Structured Materials
ISBN 978-81-322-2630-7   ISBN 978-81-322-2632-1   (eBook)
DOI 10.1007/978-81-322-2632-1

Library of Congress Control Number: 2015950051

Springer New Delhi Heidelberg New York Dordrecht London

Printed on acid-free paper

Springer (India) Pvt. Ltd. is part of Springer Science+Business Media (www.springer.com)

Preface

The book paves a new dimension in electronics with the invention of a new material *tuPOY*, which changes our perception of developing electronics. Evolving on a relatively underplayed phenomenon of static electricity in scientific exploration and application, *tuPOY* upholds the potential to rival both silicon and metals as electronics of the future. Devices made of *tuPOY* present a new emblem to the technological world, where we could envision our electronic paraphernalia from a completely different perspective. A computer the size of a big wall, which could be neatly folded and kept in our pockets when not in use and laundered on a regular basis, can be imagined possible with this invention. The concept, manufacturing process, physics, and uses of *tuPOY* as the next generation material of electronics is described in this book.

Over 2600 years ago, static electricity first found mention in the books of Thales when he accidently rubbed amber with a piece of fur, and observed that it could attract lightweight objects to itself. Over the years it was conjecturized as 'charge' transferred from one object to another during the process of rubbing so that they could exert a physical force at a distance. During early mornings or mild winters, the 'static cling' we feel when we are garbed in synthetic clothing is associated with transition of the material to a temporary unstable state where it inherits and exhibits the properties of metallic substances at molecular level. We detour from this Thalenonian philosophy to evolve a new hypothesis to this phenomenon by claiming that the synthetic material has propensity to transient fleetingly to a thermally unstable state due to which it exhibits symptomatic properties of metals and semiconductors. The characteristics of a metal and semiconductor are confined permanently in the lattice of the *tuPOY* fiber using a proprietary formulated retardant.

A wealth of spectroscopy techniques are associated to practically and theoretically validate these metallic characteristics of sensing, radiating, and processing properties of *tuPOY*. Theoretical modeling of *tuPOY* is characterized by steady-state equations exploiting interchanges based on the lattice kinetics, which mathematizes an interchange phenomenon in *tuPOY*. The numerical manifestations characterize the inherent response of lattice, thereby spouting a gamut of operations of *tuPOY* devices.

The authors have antecedently given thought to the power requirements for the transistorized material. A symbiotic power generating unit (PGU) with *tuPOY* at its core scavenges power from thermal energy presenting a new dimension in operational power dynamics.

An assemblage of operational *tuPOY* devices (sensors, antennas, transistors, etc.) are obtruded in a milieu of computing scenarios establishing *tuPOY* as a dynamic link connecting kinematical thermodynamics to electrical ambiance.

Contents

Abstract

The mystique of nature, an age-old phenomenon experienced under specific temperaments like early winter mornings, yet scientifically unexplored, describes the ambit of static electricity. Embodied frequently in synthetic textiles as 'static cling', it associates itself with transition of the material to a temporary unstable state where it inherits and exhibits the properties of metallic substances. The inception of a neoteric material, *Thermally Unstable Partially Oriented Yarn* (*tuPOY*), that captures this unstability as a permanent endure and immutably embeds the characteristic of conduction and radiation of electricity, and logical processing in the non-metallic domain, is presented in this work. This radical tuPOY, manufactured from unsaturated polymer resin textiles, establishes a dynamic link connecting kinematical thermodynamics to electrical ambiance.

The production process of tuPOY encompassing transesterification and η polymerization is developed and articulately arrested using an innovatively formulated retardant. The retardant forms an energy barrier, alternating between the polymer chains in tuPOY preventing the polycondensation process of stabilization to complete. The retardant embedded in the reaction is characterized such that it does not react with tuPOY or alter its chemical and structural properties.

Conceptual advancement in material and electrical sciences precipitates from manipulating the sensing, radiating, and processing properties of tuPOY. Correlational investigation of the electric charge flow on tuPOY lattice through transmission and scanning electron microscopy, validates the conducting properties. Responsive stimuli and ejection quantization of hydrogen nuclei under the nuclear magnetic resonance helps characterize the thermally unstable properties of tuPOY. Emission and absorption of infrared radiation on tuPOY lattice categorizes the electromagnetic spectrum in which tuPOY exhibits radiation characteristics. Fingerprinting through infrared spectral domain by Fourier transform and micrographical analysis of X-ray diffraction on lattice validate similar morphology of tuPOY and partially oriented yarns conforming its inheritance to the family of textiles. Endurance of a higher diffraction count of the X-rays on tuPOY lattice, in comparison to a regular yarn attributes to the permanent thermal unstability and presence of an inherent charge threshold, countenancing logical operations and use as processing elements.

Theoretical modeling of tuPOY is characterized by steady-state equations exploiting interchanges based on the lattice kinetics, which mathematizes an interchange phenomenon in tuPOY. The numerical manifestations calibrate mathematically tuPOY's response to any external physical impetus like charge, heat, or energy flow.

For physical testament of the sensing properties and validation of the theoretical model, tuPOY is manifested as a sensor exploitable in a plethora of applications. A microstrip patch antenna designed by amalgamation of tuPOY, raw silk, and polynylon composites experimentally verifies the radiation properties. The conduction properties are satiated by successful use of tuPOY as charge conducting wires. A power generating unit (PGU) with tuPOY as its primary element scavenges power from thermal energy presenting a new dimension in operational power dynamics.

The unified synergetic operations of tuPOY devices are protruded through pervasive computing environment in a case study embodying a wireless body area network. Circadian variation in physiological signals reflecting progression of disease in more than 2000 subjects are evaluated on the test bed, yielding impressive results.

About the Authors

H.D. Mustafa has graduated Summa Cum Laude (University Gold Medal) in 2001, from TU, New Orleans, USA, with a Double Major in Computer Science and Computer Engineering and a PhD in Electrical Engineering. He is the first recipient of the prestigious Chancellor Fellowship in 1998 for pursuing Undergraduate Education. He was awarded the Honoris Causa, Doctor of Engineering Sciences in May 2015. Currently, he is the President R&D of Transocean Inc. and RIG Ltd. He is a fellow of the ACS and ICCA. He has published over 40 journal papers including publications in reputed international journals. He was the Associate Editor of Hydrocarbon Processing Journal from 2007 to 2009.

He is a Radical and dynamic multidisciplinary scientist and executive, leading interdisciplinary multibillion dollar R&D and Industrial projects. A multidisciplinary researcher, his interest is not bound to any specific field but spans all domains of engineering and science. He is a strategic visionary and team leader amalgamating institutional and industrial research and a skilled expert in observing and translating concepts into reality, with operational excellence in diverse environments.

H.D. Mustafa is the Chief Inventor of the Crude Oil Quality Improvement (COQI®) technology and concept, which has market capitalization of US$ 2.5 Billion as of 2014. He was awarded the Merit Recognition Award by NASA, USA, in 2005 and the President of India EMPI Award in 2006 by Government of India for his contributions to Research and Development. He was nominated and shortlisted for the prestigious Global Innovator of the Year Award for 2013. In July 2013, H.D. Mustafa achieved a World Record in India for the deepest oil exploration using the COQI® process.

Sunil H. Karamchandani obtained his doctorate degree from Department of Electrical Engineering at IIT Bombay in the area of Wireless Body Area Networks (WBAN). His contributions are recognized in Nature India, Proceedings of IEEE and generated extensive news coverage with leading tabloids. namely The Financial Express, Hindustan Times, and Lokmat. He was a technical consultant for Blue Star InfoTech providing interactive solutions for segmentation of abdominal aortic aneurysms in collaboration with BARCO, U.K. He was twice the recipient of

Microsoft Travel Award in 2010 and 2012. Currently he is associated with Department of Electronics and Telecommunication, D. J. Sanghvi college of Engineering, University of Mumbai and is the IDC coordinator responsible for development of innovative ideas in association with Department of Science & Technology (DST) and National Entrepreneurship Network (NEN). He has over fifty prominent publications in journals and conferences of repute and has authored two books on Applied Mathematics over a teaching carieer spanning more than ten years. He is an avid international traveler, more recently having delivered a talk on "Promising textiles, celebrating the Silicon of the Future" at "Aurel Vlaicu," University of Arad, Romania.

Shabbir N. Merchant received his B. Tech, M. Tech, and PhD degrees all from Department of Electrical Engineering, Indian Institute of Technology Bombay, India. Currently he is a Professor in Department of Electrical Engineering at IIT Bombay. He has more than 30 years of experience in teaching and research. Dr. Merchant has made significant contributions in the field of signal processing and its applications. His noteworthy contributions have been in solving state of the art signal and image processing problems faced by Indian defence. His broad area of research interests are wireless communications, wireless sensor networks, signal processing, multimedia communication, and image processing and has published extensively in these areas. He is a co-author with his students who have won Best Paper Awards. He has been a chief investigator for a number of sponsored and consultancy projects. He has served as a consultant to both private industries and defence organizations. He is a Fellow of IETE. He is a recipient of 10th IETE Prof. S. V. C. Aiya Memorial Award for his contribution in the field of detection and tracking. He is also a recipient of 9th IETE SVC Aiya Memorial Award for 'Excellence in Telecom Education'. He is a winner of the 2013 VASVIK Award in the category of Electrical & Electronic Sciences & Technology.

Uday B. Desai received the B. Tech. degree from Indian Institute of Technology, Kanpur, India, in 1974, the M.S. degree from the State University of New York, Buffalo, in 1976, and the Ph.D. degree from The Johns Hopkins University, Baltimore, U.S.A., in 1979, all in Electrical Engineering. Since June 2009 he is the Director of IIT Hyderabad.

From 1979 to 1987 he was with the Electrical Engineering Dept. at Washington State University, Pullman, WA, U.S.A. From 1987 to May 2009 he was a Professor in the Electrical Engineering Department at the Indian Institute of Technology - Bombay. He has held Visiting Associate Professor's position at Arizona State University, Purdue University, and Stanford University. He was a visiting Professor at EPFL, Lausanne during the summer of 2002. From July 2002 to June 2004 he was the Director of HP-IITM R and D Lab. at IIT-Madras.

His research interests are in wireless communication, IoT and statistical signal processing. He is the Editor of the book "Modeling and Applications of Stochastic Processes" (Kluwer Academic Press, Boston, U.S.A. 1986). He is also a co-author of 4 books dealing with signal processing and wireless communication.

Dr. Desai is a senior member of IEEE, a Fellow of INSA (Indian National Science Academy), Fellow of Indian National Academy of Engineering (INAE). He is on the board of Tata Communications Limited. He was also on the Visitation Panel for University of Ghana.

Chapter 1
Introduction

Abstract Centuries ago a philosopher in Greece discovered static electricity. Little did he know that in this advanced age, this concept would achieve a new identity! Evolving on the basic concept of static electricity, a purely textile material tuPOY, having properties of metals is established in this chapter giving rise to a new dimension in the world of electronics. A piece of cloth belonging to a nonmetallic domain, through an innovative process translates to a material exhibiting metallic properties. Camouflaged textiles, metallic polymers, carbon nanotubes, and graphene composing the current state of art are compared and contrasted in performance and properties with tuPOY. With its innovative properties of metal and its ability to form a semiconductor, tuPOY emerges as a material holding potential to revolutionize the technological domain of the future.

Keywords tuPOY · Textiles · Polymers · Graphene · Carbon nanotubes

During 600 BCE, a greek mathematician Thales, created static electricity by polishing amber with a piece of wool or fur. Over the years, the theoretical foundation behind this phenomenon was deciphered, that when materials, such as glass and wool come in contact with each other, the synthetic textile develops a negative charge and the glass rod gains a positive charge. For centuries thereon, investigators have exploited the notion of static electricity for a plenitude of applications in fields of engineering and fundamental sciences. However till the present day, scientists believe that it is simply the contact between two different materials and the transfer of free charge that causes the static electricity to be generated in the synthetic material.

In this book, a new hypothesis to this phenomenon is evolved, by claiming that the synthetic material has a propensity to transient fleetingly to a thermally unstable state, due to which it exhibits this characteristic property of conduction and generation of static electricity immutably.

An exhaustive study into the manufacturing processes of man-made fibers, such as raw silk, nylon-66, rayon, and polyester reveals their explicit property of thermal unstability at a particular stage during the production process. In the family of synthetic fibers, partially oriented yarn emerges as one of the most commonly available and easily manufactured synthetic textile. It is nonmetallic with good abrasion resistance, drapability and ability to withstand bending stresses. Partially oriented yarn

H.D. Mustafa et al., *tuPOY: Thermally Unstable Partially Oriented Yarns*,
Advanced Structured Materials 23, DOI 10.1007/978-81-322-2632-1_1

being extensively explored in the literature; control can be exercised over each and every step in its production process.

Existing production of partially oriented yarn, entails a hot molten polymer which is subjected to a three stage polycondensation process, mainly aiming at the thermal stability of the product. This dynamic fabrication process of esterification and polymerization is articulately arrested using an innovatively formulated retardant. The retardant forms an energy barrier, alternating between the polymer chains, thus preventing the breaking or forming of any further bonds. A material is developed which retains all the textile properties of partially oriented yarn along with added properties of metallic substances. A conduction property is embedded in the textile that makes it behave as a metallic substance by altering its chemical properties. The material so obtained not only exhibits properties of conduction and generation of static electricity permanently and perpetually but also retains its characteristics of. The book aims to introduce this new invention as **Thermally Unstable Partially Oriented Yarns (tuPOY)**.

Parametric verification of the production process is achieved by temperature and pressure changes measured concurrently at in the manufacturing process. Spectroscopic techniques are perpetrated for physical and chemical characterization and to explain the conformal and electronic structural information of tuPOY. Spectrometric data ratifies the textile property of tuPOY, and asserts that the material has conducting and radiating properties.

Symbolically tuPOY is mathematized by characterizing the thermodynamic kinematic changes it exhibits on embedding external energy on its lattice. This steady-state representation, embodied as an interchange phenomenon deciphers the quantitative assessment of electromagnetic theory of tuPOY, enabling its use in electrical engineering applications.

The response of tuPOY enables quantization of any physical stimulus, such as heat, light, or pressure. This property can be used for sensing any external incident energy, permitting the use of tuPOY in a multitude of applications.

Furthermore, to understand the radiating properties of tuPOY, a novel microstrip patch antenna made out of specially formulated tuPOY and synthetic textiles such as raw silk and polynylon is conceived. With our innovative formulation the required antenna size and thickness is considerably reduced in comparison to the existing state of art, and is smaller in comparison to the metallic configurations also, let alone the textile patches.

The formalistic theories illustrated in the aforementioned extracts are ascertained by practical experimentations in a pervasive environment. The operation of tuPOY as a sensor and an antenna substrate is discussed against a backdrop of pervasive healthcare monitoring scenario. tuPOY forms heart of the designed (WBAN) architecture. The waveform selected for investigation on the WBAN architecture is the radial artery. The pulses felt from the radial artery are palpable and commonly used to assess the heart rate and cardiac rhythm of the subject. The proliferation of electrical and thermal signals concomitantly between the sensors and antenna is attained by specially formulated conductive fabric fibers made of tuPOY, acting as conducting wires. The signals transmitted from the on-body antenna are received at a distant health monitoring station. Robust classification and inferencing of these signals are attained using an automated physician machine (APM), which provides an automated diagnostics and analytics, without round-the-clock presence of a physician. The power generating unit of WBAN scenario is innovatively designed using tuPOY, delineating power by absorption of heat emitted from the subject's body.

1.1 Alternative Materials as Sensors, Antennas, and Processors

The development of new sensing and radiating systems arises from the demand of highly accurate, faster and cost effective measurements. The development of sensors is inherently multidisciplinary and leverages expertise from a myriad of specialties including biology, chemistry, physics, and engineering sciences. The textiles, fibers, polymers, and threads disguised as metallic compositions which have been adapted for conducting, sensing, and radiating applications are discussed in the sections below.

1.1.1 Textile Materials

Although sensors of variety of types are well established in process industries, agriculture, medicine, and several other arenas, the development of sensing material with high sensing capabilities is proceeding at an unprecedented rate [1]. Knitted textile fabrics [2] use fine wire metallic threads like copper, due to which they generate residual inductance. The conducting surfaces being small result in a bad contact inbetween the knitted wires reducing their durability considerably. Aramid and Aracon fibers as developed by DuPont have been explored as wearable textiles [3, 4]. However, they are doped with metallic alloys, unable to absorb moisture, difficult to cut, and have low compressive strength making them an inapt choice for radiating materials or sensors.

Polypyrrole (PPy) coated textile and foam-based sensors, find their use in wearable sensing applications [5–8]. These applications generally use knitted textiles, resulting in a sensor with the ability to respond to stretch with increased conductance. However PPy sensors have prolonged response time and show variation in sensor resistance [9], usually taking several minutes before the sensor responds to any kind of stimulus. Moreover, the daintiness of PPy restricts its use in sensors, thus, its structural properties have to be improved by amalgamation with different polymers for its applications as synthetic fibers [10]. These sensors integrated with the apparels rely on the ability of its garment to stretch and thus are constructed from extensible textiles. Such a configuration relies on minimizing the degree of wearing ease in the garment [11].

Existing literature investigates extensible textiles treated with metals as antenna substrates. The claim to textile antennas [12] rests on the basis of highly conductive metalized fabric where an amalgamation of three layers of nickel, copper, and silver provide the required high conductivity. As suggested in [2] woven and non woven textiles are unfeasible as radiating materials owing to their irregular structures. Fleece textile as radiating substrate is used in [13] albeit with an aperture coupled *Shieldit* fabric which is a nylon patch coated with copper and plated with tin. The radiating patch used in the antenna is of a size five times greater than ground plane. The authors [14] employ songket, a hand woven silk or cotton patterned fabric as substrate but the conductive parts are made out of copper. The copper tape does not fasten properly, evenly, and stoutly on the songket fabric due to yielding and soft fabric substrate. Chappell et al. [15] exploit the high frequency properties of electro textile on a Duroid substrate and obtain a radiating efficiency of 78 %. Satin 5 asserted as a textile radiating patch on a Duroid substrate, involves 80 % conductive metallic threads interwoven with nonconducting material on a laminated copper ground plane of over 100 cm. A sufficiently thick felt material [16] is employed as substrate with knitted fabric patch for antenna radiations. The point of contention is that the fabric consists of entirely silver plated (16.3 nm) polyamide fibers in spite of which the antenna efficiency is lesser than 35 % and an ordinary gain of 4.4 dBi is achieved.

Textile antennas are developed in varied designs as microstrip, monopole, and planar inverted [17–19]. While these antennas are mainly metallic-doped textile materials their design is also limited as a single layer coaxial feed. A major drawback of these antennas are that they produce parasitic effects in the feed line often causing short circuits and operational interruptions. The aperture coupled patch antennas (ACPA) as proposed in [13] overcome the parasitic effects by shielding the feed lines. However, these antennas are highly susceptible to bending, affected by moisture and are prone to contamination by foreign bodies that are used for shielding. Additionally due to absence of inherent electromagnetic properties, these textile patch antennas only achieve a meager bandwidth of 2–4 % even if a large ground plane is used. Furthermore, they have low specific absorption rates due to their large size [20]. This drawback is overcome in the design of quarter wavelength monopole antennas which provides the right polarization. However, these antennas pose a major operational problem of link loss owing to their placement position. The planar inverted 'F' (PIFA) antennas as proposed in [21] provide a return loss of lower than −10 dB resulting in a very low bandwidth of 4–12 %.

1.1.2 Polymers

Polymers are prospective materials for sensing and radiating elements due to their intrinsic properties of durability, tensile strength, and high thermal resistance characteristics. Conjugated polymers are semiconductors which show metallic properties on heavy doping. Conjugated polymers exhibit electronic properties due to presence of chain of overlapping π orbitals. This is due to the presence of alternating single and double carbon-carbon ($C - C$) bonds. Electrical conductivity of conjugated polymers can be reliably regulated over a wide range through interactions with electron acceptors and donors. The interface of bonds in conjugated polymers generate influences to produce electron delocalization. The electrons contained within each orbital extend over several adjacent atoms. Due to this extension the molecular orbital is comparatively weaker which allows higher affinity with electron acceptors and donors. This movement of electrons alters the mobility and density of charge carriers considerably, leading to significant variations in polymer conductivity [22]. However, conjugated conducting polymers such as polyacetylene, polypyrrole and polyaniline as shown by [23] are ruled out due to their encapsulation requirements when operating in open air conditions and sluggish response time. The initial redox composition of the polymer electrode controls the responses to small changes in ionic concentration when the conjugated polymers are used as sensors.

Further conjugated polymers are environmentally and chemically unstable, have poor processability [24] and exhibit [25] complex film structure which is difficult to characterize. To evade the instability problem associated with conjugated polymers, conducting polymers that possess electro active components for signal transduction have been used in [22].

Conducting polymers which are intrinsically conducting, generally show highly reversible redox behavior with a distinguishable chemical memory and hence have been considered as ideal materials for the fabrication of industrial sensors [26]. Though conducting polymers have recently gained popularity as valuable sensing materials for variousorganic vapors, and as fabricating material for sensor devices, because of their applicability at room temperature, they do show changes in resistivity on exposure to different gases and humidity [27–29]. They decompose and thus have a lower shelf life. Moreover, the properties of conducting polymers depend strongly on the doping level, ion size of dopant, protonation level, and water content which have to be appropriately weighted for the sensor design [30]. Polyurethane composites are employed in form of smart foam for pressure sensing [31], nonetheless their measurements based on the variation in the conductivity of the polymer involve conjoin of copper plates.

Synthetically manufactured conductive polymers exhibit the properties of conduction but are unable to attain electromagnetic radiation characteristics which prevents their uses as antennas. The polymers that are electroactive in nature require very high operational voltage and have inherent electrostriction effects and hence cannot be seamlessly applied in device fabrication. Furthermore, they entail a complex manufacturing process and hence donot offer flexibility in their fabrication [32]. Polymers like Poly (Vinylidene Fluoride) PVDF or its copolymers P (VDF-TrFE) have piezoelectric actuation but the obtained piezoelectric constant is extremely low hence significantly limiting its applications. The circularly polarized antenna explored by [8] consists of polyamide spacer fabric with a nickel-plated woven textile as conductive material on either side. Polyamide and liquid crystalline polymers (LCP) are used as flexible substrates, however, they are basically foils and therefore, they lack drapability.

1.1.3 Graphene and Carbon Nanotubes

Advancements in nanotechnology in the last decade, have heightened the interest of graphene and carbon nanotubes as sensing, radiating, and core elements of next generation processors. Graphene is structured as a solitary layer of carbon atoms, arranged together in a single atom thick hexagonal lattice. This material with high carrier mobility at room temperature, conducts electrons even in its off state due to low current ratios of approximately 100 [33]. A Carbon Nanotube (CNT) is formed by rolling a piece of graphene ribbon to form a perfect cylinder. They display metallic or semiconducting, electron transport properties depending on the arrangement of graphitic rings in their walls [34].

Comparing their general and structural properties, it is observed that the low dimensional geometry of carbon nanotubes results in formation of kinks, leading to buckling and collapse under higher strains as is the case with graphene [35].

The strong Van Der Waals forces keep nanotubes aggregated in mesh, due to which the growth of carbon nanotubes at macroscopic lengths is intricate [36]. Chirality of the nanotube or the manner in which the carbon atoms are rolled, also controls the speed of its growth [37]. The alignment of these carbon atoms determines the electrical properties of the nanotubes. However, minuscule differences in the chiral angle cause the differences in the properties of the carbon nanotubes [38]. Also no solvent exists in which nanotubes are extremely insoluble [39].

The principal disadvantage faced by nanotubes is that they are toxic and possess health risks. The aftermath of carbon nanotubes on human health and environment have not yet been known and thus no standards have been established for judicious use of carbon nanotubes as either sensors or processing elements [40]. Carbon nanotubes fill the air passages of the lungs which may lead to suffocation. Carbon has always been associated, with dermatosis, a form of skin inflammation and results in loss of human cell viability [41]. Donaldson et al. [42] demonstrates that carbon nanotubes create asbestos like when implanted in the abdominal cavity of mice which may lead to cancer. Finally, when the logistics are concerned, Ramanathan et al. [43] report that graphene is obtained in 'bags for dollars a pound,' while single-walled nanotubes are 'hundreds of dollars per gram'.

Graphene when used as a single layer has no energy gap between the conducting and valence bands. While this gap does exists in bilayer graphene, it becomes difficult to make on/off transient changes in such devices. Researchers [44, 45] have tried to induce a significant energy gap in bilayer graphene by the application of a perpendicular electric field; and in multilayer graphene by changing the manner in which the layers are stacked above one another. Though results show that altering the stacking pattern in multilayer graphene, determines the conductivity with a tunable band gap, the fabrication of graphene structures with the aforementioned properties still represents a challenge.

During fabrication, large graphene sheets become sticky and get attached with each other [46]. This requires that polymers be wrapped around them, which increases the thickness of the graphene layers. Also, due to its high diffusivity, Graphene exists as loosely bound particles for only a short time and agglomerates rapidly. Agglomeration lowers the surface area of graphene, close to that of graphite and hence lowers its conductivity [47]. Graphene thus has its limitations when used as a sensing or processing element.

The authors [48] identify that due to the low absorption rate of 2.3 %, Graphene provides radiation properties limited to the THz range. The technology advance in THz is minified, as THz wave transceivers are not as well developed [49] compared to their microwave and optical correlates. This restriction on the operating frequency in the antenna design, thwarts, the applications of array antennas over graphene substrate in a variety of wireless applications.

Microstrip patch antenna proposed by Dragoman et al. [48] constitute graphene layer grown on 300 nm SiO_2 substrate. Graphene layer is deposited with a 1 µm. gold film to produce a gain of 2.72 dBi and efficiency of 95 % at frequency of 1 THz. Gold films are used in the fabrication process of antenna arrays to chip off the graphene patterns formed on the source, so that they can be printed onto the substrates [50]. Dragoman observed that to obtain a resonance at 120 GHz, gold film deposition of 2 µm is required, which provides a lower gain of 2.18 dBi with relatively similar radiation efficiency. Obtaining resonance in lower GHz range, requires an increase in the deposition thickness of the gold film with stagnated value of gain and efficiency. The increase in thickness of the film is constrained by the dimensions of the antenna. Thus, a different assembly line is required during the fabrication process for maintaining the specifications of the developed antennas. The researchers [51] show that for accurate resonance and size reduction, the lowest frequency that can be suitable for a CNT antenna is nearly 100 GHz. This is negligible in the frequency range from 100 to 1000 GHz but increases with the decrease in frequency. However below 100 GHz, the inductive effect due to the delayed reflected signal does not compensate completely the capacitive effect of the dipole arms. The attenuation coefficient, produces a damping effect and the resonance can never be achieved due to the non-nullification of capacitive and inductive components. The wave propagating on the arms of the dipole is highly attenuated with the factor becoming more pronounced in the frequency range of 1−10 GHz.

For a single CNT dipole antenna, the gain is very low of about 2 dB due to their extremely small radius. Moreover the average density of CNTs (10^9/cm^2) is large and simulation of such an environment is impracticable. Hence, [52] have modeled CNT antennas as N × N array elements. However, their results show that for 5 × 5, 15 × 15, 25 × 25 CNT arrays the maximum gain is 5.92, 8.19 and 8.50 dB, respectively. Thus dimension scaling, does not provide justice to the CNT properties, as the percentage bandwidth and gain obtained is sufficiently low as compared to the manufacturing costs and design limitations involved. In addition, as noted from the above results, the gain of the antenna can be only increased by increasing the size of the antenna array.

Mehdipur et al. [53] develop fractal antennas using nanotubes. The film of the single wall carbon nanotube (SWCN) undergoes a hardening process since it is a soft and flexible material. The process of hardening causes the dc conductivity of the material to be reduced by about 75 %. Moreover the process of substrate preparation is elaborate involving resin infiltration. The reduction in dc conductivity causes the increase in the resistance, which results in lower gain (4.44 dB at 5.7 GHz) and reduced radiation efficiency.

All the king's men, as we refer to the metallurgic textiles, electroactive, conjugated and conducting polymers and the buckytubes (carbon nanotubes), are inefficacious in advocating a completely nonmetallic blueprint. This book traces the creation of a thermally unstable synthetically manufactured purely textile polymer, tuPOY, having conducting, sensing and radiating properties, which overcomes all the drawbacks enlisted above; presenting a new dimension to electronics of future.

The naturally occurring phenomenon of static electricity is quantified and modeled in scientific domain as tuPOY, a first of its kind unsaturated resin polymer textile. Arresting permanently into a thermally unstable state with an innovatively formulated retardant, the neoteric material establishes a dynamic link between kinematical thermodynamics and electrical ambiance. The detailed manufacturing process for tuPOY is conceptually and experimentally outlined in this book.

tuPOY as a surrogate to metallic elements, is evidenced using an assortment of spectroscopic techniques in conventional and nonconventional investigations. Electron microscopy validates the conduction properties while X-ray based confirm the textile nature and thermal unstability. Magnetic resonance in the nuclear domain quantifies this thermal unstability and verifies the respective theoretical claims. Electromagnetic radiation is visualized in tuPOY through its interaction with infrared spectrum in the transform domain, which also establishes its molecular fingerprinting.

Theoretical modeling is characterized by steady state equations exploiting interchanges based on the lattice kinetics, which mathematizes an Interchange Phenomenon in tuPOY. The numerical manifestations calibrate mathematically, tuPOY's response to any external physical impetus like charge, heat or energy flow delineating a new scientific arena.

Antiphony of tuPOY to external stimuli characterized by the aforementioned mathematical domain as an inherent microstress, outlines its use as sensors. The microstress is also allegorical to the threshold in metal oxide semiconductor transistor gates, beyond which the device starts conducting. This work innovatively explores the varied uses of this internal threshold in the material. The conduction properties are satiated by successful use of tuPOY as charge conducting wires. A power generating unit with tuPOY as its primary element, scavenges power from thermal energy, presenting a new dimension in operational power dynamics. Meditating a composite textile antenna with tuPOY, raw silk, and polynylon proves its electromagnetic radiation proficiency.

The proposed new electronics are tested under a judiciously designed case study, comprising of more than 2000 volunteers, in a pervasive computing environment. A wireless body area network is designed and tested on radial pulse artery for varied diseases of heart, liver, and lungs. For analytics, an automated physician module comprising of an artificial intelligence core, is embedded in the back-end of the architecture. The proposed case study yields impressive results ably supported by a robust architecture with minimal tradeoffs.

A phenomenon [54–58] is observed in nature, in which synthetic clothing, sometimes under very specific conditions, generate and conducts an electric current on its surface. This phenomenon [59–61] is observed mainly during winters or early mornings and lasts only for a small duration of time. An important observation however, is that this textile material induces itself to a temporary unstable state where it inherits and exhibits properties of a metal. This work identifies the stage in the production process of partially oriented yarns, where it attains unstability and captures this stage in a permanent condition. The material so obtained exhibits conduction and radiation properties that are tested in this manuscript with proposed novel textile sensors and antennas. Current trends involve smart materials, which conduct and generate electricity when doped with metallic formulations or threads. The following scenarios represent a comprehensive overview of the claims of inherent conductive textiles and polymers either as sensors or antennas. Even though majority of state of art claims to be in a nonmetallic dimension, our exhaustive review reveals they camouflage a falsity.

References

1. Parvatikar, N., Jain, S., Khasim, S., Revansiddappa, M., Bhoraskar, S., Ambika Prasad, M.N.: Electrical and humidity sensing properties of polyaniline/WO3 composites. Sens. Actuators B: Chem. **114**(2), 599–603 (2006)
2. Massey, P.J.: Mobile phone fabric antennas integrated within clothing. In: Eleventh International Conference on Antennas and Propagation (IEE Conf. Publ. No. 480), vol. 1, pp. 344–347 (2001)
3. Declercq, F., Rogier, H.: Active integrated wearable textile antenna with optimized noise characteristics. IEEE Trans. Antennas Propag. **58**(9), 3050–3054 (2010)
4. Winterhalter, C., Teverovsky, J., Wilson, P., Slade, J., Farell, B., Horowitz, W., Tierney, E.: Development of electronic textiles for U.S. military protective clothing systems. Stud. Health Technol. Inform. **108**, 194–198 (2004)
5. Farringdon, J., Moore, A., Tilbury, N., Church, J., Biemond, P.: Wearable sensor badge and sensor jacket for context awareness. In: The Third International Symposium on Wearable Computers: Digest of Papers, pp. 107–113 (1999)
6. Lorussi, F., Rocchia, W., Scilingo, E., Tognetti, A., De Rossi, D.: Wearable, redundant fabric based sensor arrays for reconstruction of body segment posture. IEEE Sens. J. **4**(6), 807–818 (2004)
7. Mazzoldi, A., Rossi, D.D., Lorussi, F., Scilingo, E.P., Paradiso, R.: Smart textiles for wearable motion capture systems. AUTEX Res. J. **2**(4), 199–203 (2002)
8. Troster, G.: The agenda of wearable healthcare. IMIA Yearbook of Medical Informatics 2005: Ubiquitous Health Care Systems, pp. 125–138 (2005), cited By (since 1996) 22
9. Engin, M., Demirel, A., Engin, E.Z., Fedakar, M.: Recent developments and trends in biomedical sensors. Measurement **37**(2), 173–188 (2005). http://www.sciencedirect.com/science/article/pii/S0263224104001113
10. Xue, P., Tao, X., Kwok, K.W., Leung, M., Yu, T.: Electromechanical behavior of fibers coated with an electrically conductive polymer. Text. Res. J. **74**(10), 929–936 (2004). http://trj.sagepub.com/content/74/10/929
11. Dunne, L., Brady, S., Smyth, B., Diamond, D.: Initial development and testing of a novel foam-based pressure sensor for wearable sensing. J. NeuroEng. Rehabil. **2**(1), 4 (2005). http://www.jneuroengrehab.com/content/2/1/4

12. Klemm, M., Troester, G.: Textile UWB antennas for wireless body area networks. IEEE Trans. Antennas Propag. **54**(11), 3192–3197 (2006)
13. Hertleer, C., Tronquo, A., Rogier, H., Vallozzi, L., Van Langenhove, L.: Aperture-coupled patch antenna for integration into wearable textile systems. IEEE Antennas Wirel. Propag. Lett. **6**, 392–395 (2007)
14. Rahman Osman, M.A., Bin Rahim, M.K.: Wearable textile antenna: fabrics investigation. J. Commun. Comput. **7**(7), 75–80 (2010)
15. Ouyang, Y., Chappell, W.: High frequency properties of electro-textiles for wearable antenna applications. IEEE Trans. Antennas Propag. **56**(2), 381–389 (2008)
16. Locher, I., Klemm, M., Kirstein, T., Troster, G.: Design and characterization of purely textile patch antennas. IEEE Trans. Adv. Packag. **29**(4), 777–788 (2006)
17. Salonen, P., Yang, F., Rahmat Samii, Y., Kivikoski, M.: WEBGA—wearable electromagnetic band-gap antenna. In: Antennas and Propagation Society International Symposium, vol. 1, pp. 451–454. IEEE (2004)
18. Tronquo, A., Rogier, H., Hertleer, C., Van Langenhove, L.: Robust planar textile antenna for wireless body lans operating in 2.45 GHz ISM band. Electron. Lett. **42**(3), 142–143 (2006)
19. Ouyang, Y., Karayianni, E., Chappell, W.: Effect of fabric patterns on electrotextile patch antennas. In: Antennas and Propagation Society International Symposium, vol. 2B. pp. 246–249. IEEE (2005)
20. Park, H., Choi, J.: Design of broad quad-band planar inverted-F antenna for cellular/PCS/UMTS/DMB applications. Microw. Opt. Technol. Lett. **47**(5), 418–421 (2005). http://dx.doi.org/10.1002/mop.21188
21. Chen, C.C., Volakis, J.L.: Bandwidth broadening of patch antennas using nonuniform substrates. Microw. Opt. Technol. Lett. **47**(5), 421–423 (2005). http://dx.doi.org/10.1002/mop.21189
22. Dai, L., Soundarrajan, P., Kim, T.: Sensors and sensor arrays based on conjugated polymers and carbon nanotubes. Pure Appl. Chem. **74**(9), 1753–1772 (2002)
23. Harun, M.H., Saion, E., Kassim, A., Yahya, N., Mahmud, E.: Conjugated conducting polymers: a brief overview. Sens. Peterb. NH **2**, 63–68 (2007). http://sedaya.edu.my/jasa/2/papers/08I.pdf
24. Bhakshi, A., Bhalla, B.: Electrically conducting polymers: materials of the twenty first century. J. Sci. Ind. Res. **63**(9), 392–395 (2004)
25. McNeill, C.: Organic devices. New perspectives provided from soft X-ray characterization. In: APS Meeting Abstracts, p. 41001 (2011)
26. Dhawan, S., Kumar, D., Ram, M., Chandra, S., Trivedi, D.: Application of conducting polyaniline as sensor material for ammonia. Sens. Actuators B: Chem. **40**(2–3), 99–103 (1997). http://www.sciencedirect.com/science/article/pii/S092540059780247X
27. Bartlett, P.N., Ling-Chung, S.K.: Conducting polymer gas sensors part III: results for four different polymers and five different vapours. Sens. Actuators **20**(3), 287–292 (1989). http://www.sciencedirect.com/science/article/pii/0250687489801271
28. Agbor, N., Petty, M., Monkman, A.: Polyaniline thin films for gas sensing. Sens. Actuators B: Chem. **28**(3), 173–179 (1995). http://www.sciencedirect.com/science/article/pii/0925400595017259
29. Barker, P., Chen, J., Agbor, N., Monkman, A., Mars, P., Petty, M.: Vapour recognition using organic films and artificial neural networks. Sens. Actuators B: Chem. **17**(2), 143–147 (1994). http://www.sciencedirect.com/science/article/pii/092540059487042X
30. Gök, A., Sari, B., Talu, M.: Conducting polyaniline sensors for some organic and inorganic solvents. Int. J. Polym. Anal. Charact. **11**(3), 227–238 (2006). http://www.tandfonline.com/doi/abs/10.1080/10236660600678953
31. Brady, S., Diamond, D., Lau, K.T.: Inherently conducting polymer modified polyurethane smart foam for pressure sensing. Sens. Actuators A: Phys. **119**(2), 398–404 (2005). http://www.sciencedirect.com/science/article/pii/S092442470400771X

32. Mazzone, A., Zhang, R., Kunz, A.: Novel actuators for haptic displays based on electroactive polymers. In: Proceedings of the ACM Symposium on Virtual Reality Software and Technology, ser. VRST'03. New York, USA: ACM, pp. 196–204 (2003). http://doi.acm.org/10.1145/1008653.1008688

33. Xia, F., Farmer, D.B., Lin, Y.M., Avouris, P.: Graphene field effect transistors with high on/off current ratio and large transport band gap at room temperature. Nano Lett. **10**(2), 715–718 (2010) pMID: 20092332. http://pubs.acs.org/doi/abs/10.1021/nl9039636

34. Odom, T.W., Huang, J.L., Kim, P., Lieber, C.M.: Atomic structure and electronic properties of single-walled carbon nanotubes. Nature **391**(6662), 62–64 (1998). http://dx.doi.org/10.1038/34145

35. Xu, Z., Buehler, M.J.: Strain controlled thermomutability of single-walled carbon nanotubes. Nanotechnology **20**(18), 185701 (2009). http://www.ncbi.nlm.nih.gov/pubmed/19420624

36. Dresselhaus, M.S., Dresselhaus, G., Avouris, P.: Carbon nanotubes: synthesis, structure, properties, and applications. Topics in Applied Physics, vol. 80, pp. 29–53. Springer, Berlin (2001)

37. Wernik, J.M., Meguid, S.A.: Recent developments in multifunctional nanocomposites using carbon nanotubes. Appl. Mech. Rev. **63**(5), 050801 (2010). http://link.aip.org/link/?AMR/63/050801/1

38. Hassanien, A., Tokumoto, M., Kumazawa, Y., Kataura, H., Maniwa, Y., Suzuki, S., Achiba, Y.: Atomic structure and electronic properties of single-wall carbon nanotubes probed by scanning tunneling microscope at room temperature. Appl. Phys. Lett. **73**(26), 3839–3841 (1998). http://link.aip.org/link/?APL/73/3839/1

39. Kim, T.H., Doe, C., Kline, S.R., Choi, S.M.: Organic solvent redispersible isolated single wall carbon nanotubes coated by in-situ polymerized surfactant monolayer. Macromolecules **41**(9), 3261–3266 (2008). http://pubs.acs.org/doi/abs/10.1021/ma702684e

40. Zhao, X., Liu, R.: Recent progress and perspectives on the toxicity of carbon nanotubes at organism, organ, cell, and biomacromolecule levels. Environ. Int. **40**(0), 244–255 (2012). http://www.sciencedirect.com/science/article/pii/S0160412011002832

41. Murray, A., Kisin, E., Leonard, S., Young, S., Kommineni, C., Kagan, V., Castranova, V., Shvedova, A.: Oxidative stress and inflammatory response in dermal toxicity of single walled carbon nanotubes. Toxicology **257**(3), 161–171 (2009). http://www.sciencedirect.com/science/article/pii/S0300483X08006124

42. Poland, C., Duffin, R., Kinloch, I., Maynard, A., Wallace, W., Seaton, A., Stone, V., Brown, S., MacNee, W., Donaldson, K.: Carbon nanotubes introduced into the abdominal cavity of mice show asbestos like pathogenicity in a pilot study. Nat. Nanotechnol. **3**(7), 423–428 (2008)

43. Ramanathan, T., Abdala, A.A., Stankovich, S., Dikin, D.A., Herrera Alonso, M., Piner, R.D., Adamson, D.H., Schniepp, H.C., Chen, X., Ruoff, R.S.: Functionalized graphene sheets for polymer nanocomposites. Nat. Nanotechnol. **3**(6), 327–331 (2008). http://dx.doi.org/10.1038/nnano.2008.96

44. Lui, C.H., Li, Z., Mak, K.F., Cappelluti, E., Heinz, T.F.: Observation of an electrically tunable band gap in trilayer graphene. Nat. Phys. **7**, 944–947 (2011)

45. Wu, B.R.: Field modulation of the electronic structure of trilayer graphene. Appl. Phys. Lett. **98**(26), 263107 (2011). http://link.aip.org/link/?APL/98/263107/1

46. Zook, J.M., Rastogi, V., Maccuspie, R.I., Keene, A.M., Fagan, J.: Measuring agglomerate size distribution and dependence of localized surface plasmon resonance absorbance on gold nanoparticle agglomerate size using analytical ultracentrifugation. ACS Nano. http://dx.doi.org/10.1021/nn202645b

47. Li, D., Mueller, M.B., Gilje, S., Kaner, R.B., Wallace, G.G.: Processable aqueous dispersions of graphene nanosheets. Nat. Nanotechnol. **3**(2), 101–105 (2008). http://www.nature.com/nnano/journal/v3/n2/full/nnano.2007.451.html

48. Dragoman, M., Muller, A.A., Dragoman, D., Coccetti, F., Plana, R.: Terahertz antenna based on graphene. J. Appl. Phys. **107**(10), 104313 (2010). http://link.aip.org/link/?JAP/107/104313/1

49. Wang, Y., Wu, Q., Shi, W., He, X., Sun, X., Gui, T.: Radiation properties of carbon nanotubes antenna at terahertz/infrared range. Int. J. Infrared Millim. Waves **29**, 35–42 (2008). http://dx.doi.org/10.1007/s10762-007-9306-9

50. Song, L., Ci, L., Gao, W., Ajayan, P.M.: Transfer printing of graphene using gold film. ACS Nano **3**(6), 1353–1356 (2009) pMID: 19438194. http://pubs.acs.org/doi/abs/10.1021/nn9003082

51. Attiya, M.: Lower frequency limit of carbon nanotube antenna. Prog. Electromagn. Res. **94**, 419–433 (2009). http://link.aip.org/link/?JAP/107/104313/1

52. Shi, W., Wang, Y., Wu, Q., Wang, X.: Terahertz properties of carbon nanotubes antenna arrays. In: Zhang, C., Zhang, X.-C. (eds.) SPIE, vol. 6840, no. 1, p. 684006 (2007). http://link.aip.org/link/?PSI/6840/684006/1

53. Mehdipour, A., Rosca, I.D., Sebak, A.R., Trueman, C.W., Hoa, S.V.: Full composite fractal antenna using carbon nanotubes for multiband wireless applications. IEEE Antennas Wirel. Propag. Lett. **9**, 891–894 (2010)

54. Hench, L.: Biomaterials. Science **208**(4446), 826–831 (1980). http://www.sciencemag.org/content/208/4446/826.short

55. Calvert, P.: Polymers that make light work. Nature **337**, 408–409 (1989). http://dx.doi.org/10.1038/337408a0

56. Moore, W.R.: Adhesion and thermal degradation of high polymers. Nature **205**, 1146–1147 (1965)

57. Cheng, S.Z.D.: Materials science: polymer crystals downsized. Nature **448**, 1006–1007 (2007). http://dx.doi.org/10.1038/4481006a

58. Lemstra, P.J.: Confined polymers crystallize. Science **323**(5915), 725–726 (2009). http://www.sciencemag.org/content/323/5915/725.short

59. Ragauskas, A.J., Williams, C.K., Davison, B.H., Britovsek, G., Cairney, J., Eckert, C.A., Frederick, W.J., Hallett, J.P., Leak, D.J., Liotta, C.L., Mielenz, J.R., Murphy, R., Templer, R., Tschaplinski, T.: The path forward for biofuels and biomaterials. Science **311**(5760), 484–489 (2006). http://www.sciencemag.org/content/311/5760/484.abstract

60. Brown, A.E., Reinhart, K.A.: Polyester fiber: from its invention to its present position. Science **173**(3994), 287–293 (1971). http://www.sciencemag.org/content/173/3994/287.abstract

61. Yan, H., Chen, Z., Zheng, Y., Newman, C., Quinn, J.R., Dötz, F., Kastler, M., Facchetti, A.: A high-mobility electron-transporting polymer for printed transistors. Nature **457**(7230), 679–686 (2009). http://www.ncbi.nlm.nih.gov/pubmed/19158674

Chapter 2
Manufacturing Process of Thermally Unstable Partially Oriented Yarns

Abstract The metamorphosis of tuPOY from basic organic raw materials to an advanced conducting fiber having metallic properties is discussed in this chapter. A step by step process of embedding conduction and radiation properties in tuPOY, with an elaborated stagewise analysis is presented. The production process is justified with parametric evaluation and analysis. Production process flow from a commercial manufacturing viewpoint is outlined.

Keywords Thermal unstability · Polymerization · Transesterification · Polycondensation · Retardant · Energy barrier

A specific property of polyester based yarn is its thermal unstability at a particular stage during the production process [1–7] . The existent polymerization of POY seeks to remove the molecular unstability through a polycondensation process. In the proposed work, this instability is taken advantage of and the reaction is arrested in the polycondensation stage to obtain an unstable tuPOY. The reaction mechanism initiates with the esterification of purified Dimethyl Terephthalate (DMT) and Ethylene Glycol (EG) in a ratio of 1:3, which produces oligomer, as shown in the reactions 1–3 (Fig. 2.1).

Reaction I, includes a complex of of EG with carbomethoxy groups of DMT, methyl β-hydroxyethyl terephthalate and with oligomers containing carbomethoxy end groups. The reaction produces bis-hydroxyl ethyl terephthalate (BHET) as the main monomer. Reaction II entails a complex of transesterification reactions of carbomethoxy group with the carbo-hydroxyethoxy groups of bis-β-hydroxyethyl terephthalate (β-BHET) and all oligomers containing carbohydroxyethoxy groups. Reaction III outlines the two stage condensation reactions of the carbohydroxyethoxy group. During polycondensation process the oligomers and polymer chains combine to produce lengthened chains in efforts to stabilize their entropy. The thermally unstable reaction is arrested at this point using an innovatively proprietary retardant. The retardant forms an alternating barrier between the $C - H$, $C - C$ and $C - O$ chains preventing the breaking or forming of any further chains of oligomers. An advantage of the innovative retardant is that, it does not react with the oligomers and the end product shows no reactivity with the formulated retardant. Furthermore, the reaction is arrested in the third stage because at this stage all the properties of the textile

H.D. Mustafa et al., *tuPOY: Thermally Unstable Partially Oriented Yarns*, Advanced Structured Materials 23, DOI 10.1007/978-81-322-2632-1_2

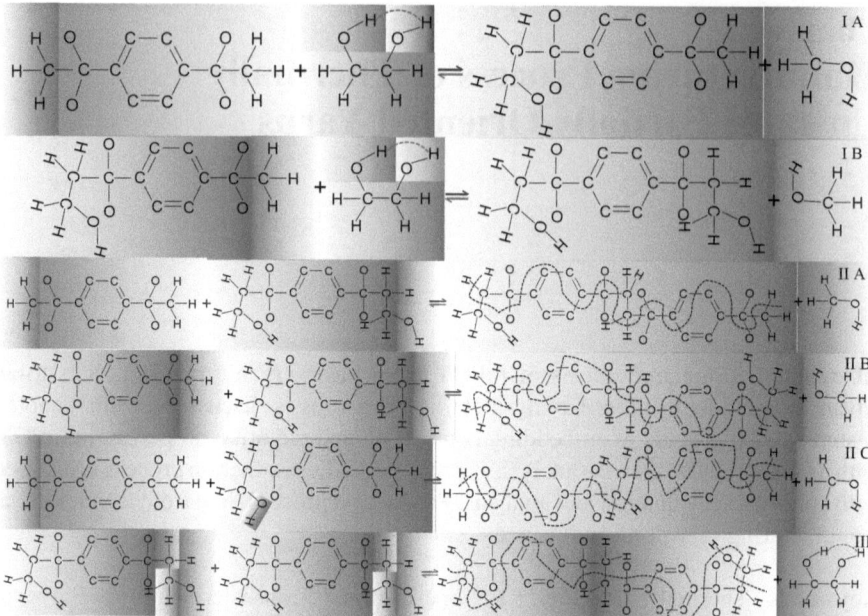

Fig. 2.1 Reaction mechanism occurring during the production of tuPOY. (1) Shows the complex esterification mechanism forming oligomers, (2a–c) shows the complex transesterification of the oligomers within themselves to form polymer chain, (3) shows the polycondensation of the transesterification process. The retardant is added on completion of (3) and the retardant forms an energy barrier which is indicated by the dashed line. The energy barrier from the retardant alternates between the $C - H$, $C - C$ and $C - O$ bonds. The end products show no reactivity between the formulated retardant and tuPOY

are attained and the reaction does not reverse towards the initial stage, resulting in Thermally Unstable Partially Oriented Yarns termed as **tuPOY**.

2.1 Production Process of tuPOY

Esterification, transesterification and polycondensation forms the quintessence for polymerization of any synthetic polymer yarn [8–10]. Esterification reactions of polyester blends occur in the molten state via alcoholysis, acidolysis and direct interchange reactions. The exchange reaction results from alcoholysis between an ester[1] and an alcohol.[2] The acidolysis reactions involve carboxyl-terminated oligomers

---

[1] The ester used in our experiments is AR grade. Dimethyl terephthalate procured from MERCK chemicals

[2] The alcohol used in our experiments is XR grade. Ethylene Glycol procured from GlaxoSmithKline

formed during esterification while the transesterification proceeds via a direct interaction between carbohydroxyethoxy groups.

Esterification of DMT is the key initiation step in the industrial production of tuPOY. The complete process flow and instrumentation diagram for tuPOY is illustrated in Fig. 2.2. The DMT briquettes are oxidized at temperature of 140 °C where the excess moisture is removed and the off gases are incinerated. The process of oxidation is followed by flash cooling and of the residue to obtain molten DMT. Esterification involves a reaction of molten DMT and EG stirred at 2000 r.p.m for over four hours. The repulsive nature of the hydrogen atoms in the alcohol initiates the esterification process at 190 °C. This repulsive nature, causes the $C - O$ bonds to break and the liberated hydroxyl group of EG combines with methoxy group of the ester as observed in reaction 1(A). This occurs because the $C - C$ bond in ester is non-covalent and weak. Esterification entails the interchange of methoxy group in ester and the ethoxy group with the exoneration of methanol during the reaction. Reaction 1(A) results in the formation of an oligomer with carbomethoxy end group, and esterification process continues until all the alcohol is consumed. A similar exchange reaction occurs as observed in reaction 1(B). The esterification reactions in 1(A) and 1(B) illustrate the transformation of each methyl ester group on the DMT molecule into a hydroethoxyethyl ester group. The reaction progresses towards the transesterification stage with oligomers containing methoxy and hydroethoxy

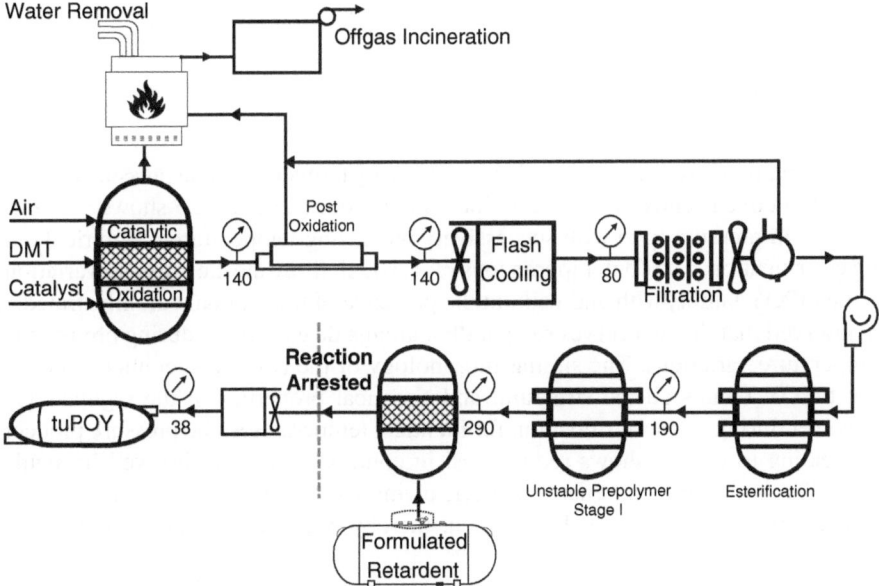

Fig. 2.2 Production process and instrumentation flow of tuPOY, showing arresting of the reaction at the polycondensation stage after transesterification. The final product is obtained at atmospheric temperature and pressure, retaining all the properties of synthetic partially oriented yarns

end groups. Esterification is ensued by transesterification of the carbomethoxy groups and condensation of the carbohydroethoxy groups. Transesterification and polycondensation reactions illustrated by reactions 2(a), 2(b) and 2(c) and 3 respectively result in the formation of linear oligomers of β-BHET family at temperatures of 290 °C. Transesterification reactions of carbohydroethoxy and carbomethoxy groups enter into 2(a) and 2(c) respectively while complex of transesterifications of carbomethoxy groups with oligomers containing carbohydroxyethoxy groups enter into reaction 2(b). Reaction 3 illustrates the complex of condensation reactions of the carbohydroxyethoxy groups. The retardant is added at the pre condensation level and reaction is allowed to arrest for about 4–5 hours. The retardant forms an energy barrier, indicated by the dashed line and the reaction is arrested at this point. The retardant prevents the linear BHET oligomers formed during transesterifications reaction to combine with each other, due to its energy barriers at the polymer chaining points. The barrier created by the retardant alternates between the bonds and the rings in BHET oligomers and hence prevents formation of any further polymer chains. The material thus arrested is in thermally unstable state and the reaction is prevented to complete the polycondensation stage. The effect of the retardant is not to change the properties of the regular POY but to stop the formation of polymeric chains which embed structural and chemical stability. The developed material is extracted at atmospheric temperature in molten form. It is then pressure extruded[3] and drawn out into fibers. The fiber produced during the process correlates with physical characteristics of raw of 78.2 [11–13].

2.2   Parametric Verification

tuPOY manufactured at the lab scale is vigilated by temperature and pressure changes measured concurrently at fiducial points in the process. Figure 2.3 shows a gradual increase in pressure (psi) with successively increasing temperatures, empirical during esterification. The contemporaneous plots which illustrate the pressure variations in the tuPOY lattice, with and without the presence of the retardant are illustrated. It is observed that the two curves run parallel, alongside each other during preliminary temperature variations. The similar morphology of the two plots, connotes the fact that tuPOY retains all the structural and chemical properties of the regular POY. However during transesterification, for identical temperatures, the pressure plot corresponding to tuPOY shows reduced psi in contrast with that observed in regular POY. The build up of an energy barrier, owing to the addition of the retardant in tuPOY, thwarts the pace of the reaction, thereby reducing the rate of increase in pressure.

---

[3]Pressure extrusion to obtain a 78.2 denier tuPOY, is performed on a lab scale E10 from Oerlikon Barmeg with a titer of 0.5–1.5 dpf. Further information can be obtained at http://www.barmag. oerlikontextile.com/desktopdefault.aspx/tabid-431/.

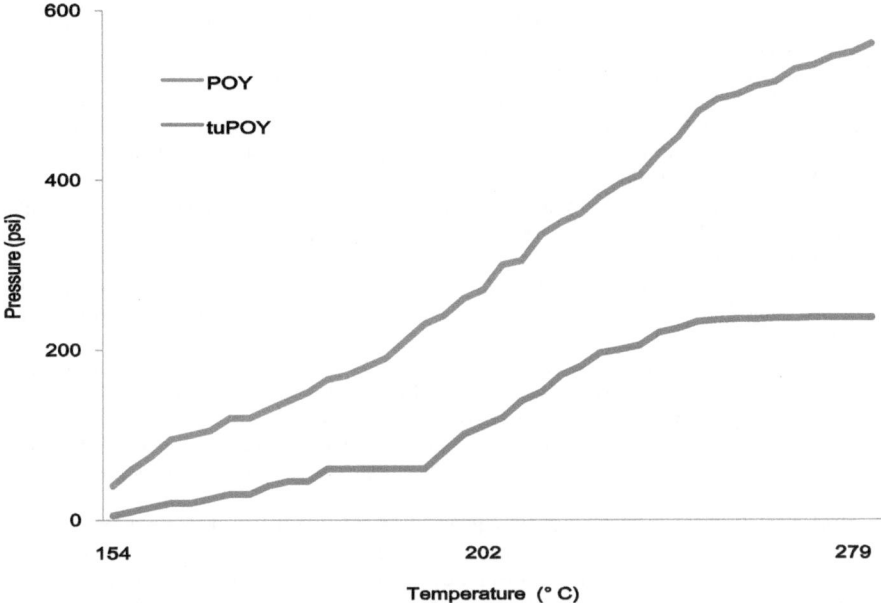

Fig. 2.3 Parametric verification in the fabrication process of tuPOY. tuPOY progressively attains a steady state due to the action of the retardant

At the onset of polycondensation, the tuPOY curve progressively attains a steady state, while in the case of the regular POY the pressure gradually increases with temperature and the graph follows its normal course. The coordinates on the graph, where the pressure tends to approach a steady state value, indicate that the reaction is arrested at this point and the material is captured in its thermally unstable state.

References

1. Calvert, P.: Polymers that make light work. Nature, **337**, 408–409 (1989). http://dx.doi.org/10.1038/337408a0
2. Moore, W.R.: Adhesion and thermal degradation of high polymers. Nature **205**, 1146–1147 (1965)
3. Cheng, S.Z.D.: Materials science: Polymer crystals downsized. Nature **448**, 1006–1007 (2007). http://dx.doi.org/10.1038/4481006a
4. Lemstra, P.J.: Confined polymers crystallize. Science **323**(5915), 725–726 (2009). http://www.sciencemag.org/content/323/5915/725.short
5. Özkan, G., Ürkmez, G., Özkan, G.: Application of Box-Wilson optimization technique to the partially oriented yarn properties. Polym. Plast. Technol. Eng. **42**(3), 459–470 (2003). http://www.scopus.com/inward/record.url?eid=2-s2.0-0042155545&partnerID=40&md5=8d844cfbe181b6ebf201aaeda2803d42 (cited By (since 1996) 5)
6. Ragauskas, A.J., Williams, C.K., Davison, B.H., Britovsek, G., Cairney, J., Eckert, C.A., Frederick, W.J., Hallett, J.P., Leak, D.J., Liotta, C.L., Mielenz, J.R., Murphy, R., Templer, R.,

Tschaplinski, T.: The path forward for biofuels and biomaterials. Science **311**(5760), 484–489 (2006). http://www.sciencemag.org/content/311/5760/484.abstract

7. Brown, A.E., Reinhart, K.A.: Polyester fiber: from its invention to its present position. Science **173**(3994), 287–293 (1971). http://www.sciencemag.org/content/173/3994/287.abstract

8. Smirnov, P.V., Repina, L.P., Bunigina, N.S., Aizenshtein, M., Kvasha, N.M., Kiselev, V.V.: Transesterification of dimethyl terephthalate with ethylene glycol. Fibre Chem. **15**, 332–336 (1984). doi:10.1007/BF00548126. http://dx.doi.org/10.1007/BF00548126

9. Ahn, Y.C., Choi, S.M.: Analysis of the esterification process for poly(ethylene terephthalate). Macromol. Res. **11**, 399–409 (2003). doi:10.1007/BF03218968. http://dx.doi.org/10.1007/BF03218968

10. Ravindranath, K., Mashelkar, R.: Polyethylene terephthalate-II. Engineering analysis. Chem. Eng. Sci. **41**(12), 2969–2987 (1986). http://www.sciencedirect.com/science/article/pii/0009250986850345

11. Yan, H., Chen, Z., Zheng, Y., Newman, C., Quinn, J.R., Dötz, F., Kastler, M., Facchetti, A.: A high-mobility electron-transporting polymer for printed transistors. Nature **457**(7230), 679–686 (2009). http://www.ncbi.nlm.nih.gov/pubmed/19158674

12. Kim, O.K., Little, R.C., Patterson, R.L., Ting, R.Y.: Polymer structures and turbulent shear stability of drag reducing solutions. Nature **250**, 408–410 (1974). http://dx.doi.org/10.1038/250408a0

13. Park, S.I., Xiong, Y., Kim, R.H., Elvikis, P., Meitl, M., Kim, D.H., Wu, J., Yoon, J., Yu, C.J., Liu, Z., Huang, Y., Hwang, K.C., Ferreira, P., Li, X., Choquette, K., Rogers, J.A.: Printed assemblies of inorganic light-emitting diodes for deformable and semitransparent displays. Science **325**(5943), 977–981 (2009). http://www.sciencemag.org/content/325/5943/977.abstract

Chapter 3
Thermal Unstability Analysis

Abstract The physics of tuPOY in relation to its molecular structure forms the essence of this chapter. The manufacturing process is experimentally and theoretically justified by a variety of spectroscopical imaging, nonimaging, and micro graphical procedures. These techniques provide us with a physical insight into the conducting and radiating behavior of tuPOY. The juxtaposition of these investigations outlines the mechanism of tuPOY as the next generation processing element.

Keywords Spectroscopy · Electron microscopy · Micro graphical analysis · X-ray diffraction · Nuclear magnetic resonance · Fourier transform infrared spectroscopy

Spectroscopy techniques are used to perform an interaction between tuPOY and its dissipated or radiated energy as a function of wavelength. Spectroscopic techniques help us to determine the nature, characteristics, lattice, and structural transformations of tuPOY. The methods employed in this work encompasses a variety of imaging techniques such as electron microscopy, micro graphical analysis such as XRD (X-ray Diffraction) and several nonimaging techniques like NMR (Nuclear Magnetic Resonance), and FTIR (Nuclear Magnetic Resonance).

X-ray spectroscopy studies, performed to determine the percentage crystallinity in the tuPOY lattice, reveal broad halo patterns derived from the amorphous chain interactions. XRD analysis helps us in validating the inherent textile properties of tuPOY. The diffraction pattern, in angle form, shows peaked reflections at Bragg angle around 21°. The molecular fingerprinting of tuPOY lattice is established with Fourier Transform Infrared spectroscopy (FTIR). FTIR confirms the morphology of the regular POY, similar to that of tuPOY. Beyond this, in infrared spectroscopy, the vibrations of the atoms in a molecule adjudicate that the material exhibits resonance on its surface in specific frequency ranges, validating the radiation characteristics of tuPOY. NMR techniques are used for material characterization by analysis of final and intermediate products, and record the intermediate alcohol ratios to test the thermal unstability of the material. Transmission electron microscopy and scanning electron imaging techniques provide an extremely useful method for the confirmation of the conducting properties of tuPOY.

In interpretation of these tests, an unconventional modus operandi is developed in their analysis to realize the conducting and radiating properties of the material.

© Springer India 2016   19
H.D. Mustafa et al., *tuPOY: Thermally Unstable Partially Oriented Yarns*,
Advanced Structured Materials 23, DOI 10.1007/978-81-322-2632-1_3

TEM/SEM are used to characterize the orientation of crystal lattice and produce images that highlight elemental contrast. Depending upon the elemental contrast of the high-resolution images produced by these investigations, a conclusion is reached on the conducting properties of tuPOY. The infrared spectrum obtained from FTIR represents the fingerprint of lattice. Apart from characteristic positions of the bands in the spectrum, required to understand the morphology of tuPOY, the relative strengths of the transmission bands are considered to predict the resonating states and hence the radiating properties of tuPOY. NMR relates the number of peaks in the spectrum and number of hydrogen atoms in the subsequent stages of lattice formation. The ratios of the areas underneath each peak, determines number of hydrogen atoms, which assist in validating the thermal unstability of the tuPOY lattice. The findings presented in this chapter are based on the averaging of the results obtained after carrying out spectroscopic experiments over 250 samples, from which, sharp contrast images are compiled.

3.1 TEM and SEM Analysis

The Transmission Electron Microscope (TEM) operates on the same basic principle as the light microscope but uses electrons instead of light [1]. The extremely low wavelength of electrons enables imaging resolutions to the order of 10^{-10} m. The electrons emitted from the source, travel through vacuum in the column of the microscope. An electromagnetic lens is used to focus these electrons into a very thin beam. During TEM[1] analysis, the monochromatic electron beam is transmitted perpendicularly through tuPOY, focused into an image and projected onto a phosphor-coated screen emitting visible light. tuPOY sample is mounted on a carbon-coated copper grid as a fiber of approximately $10\,\mu$m thickness. The interaction of the accelerated electrons with the sample helps to determine the elemental composition of the sample. The image processed with TEM is illustrated in Fig. 3.1a along side a scanned image of the tuPOY from the scanning microscope. In TEM, a beam of electrons is transmitted perpendicularly through the surface of tuPOY lattice, interacting with the material as they pass through. The image formed, on the orthogonal end from transmission, is magnified and focused onto a photographic film. For the molecular structure to be evident the electrons should traverse through the sample while emitting energy in the process. This reduction in energy is corresponded by the illumination observed on the imaging screen. A dark silhouette of the lattice is observed on the film corresponding to the position of the lattice on the grid. This indicates that none of the electrons which strike the material are transmitted through the lattice. At places where the electron beams transmitted through the material, do not reach the grid, a dark shadow is observed. Alternatively, the area on the image where the electron beam strikes the grid directly, without any interference from the lattice appears

---

[1]TEM is performed on PHILIPS CM200 microscope with a transmitted electron beam of 20 KV.

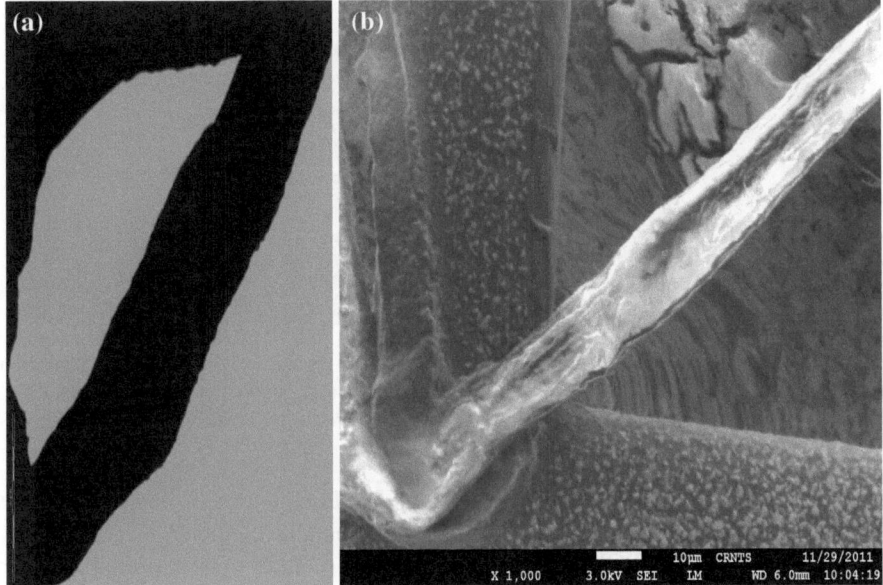

Fig. 3.1 A bright lattice is observed in scanning electron microscopy which indicates conduction properties of tuPOY which is absent in the darker lattice of TEM in Fig. 3.1a

bright and illuminated. Based on the above observation, it can be contemplated that tuPOY behaves either as a conductor or as an insulator.

This is corroborated by performing a SEM analysis,[2] where the electron beam is used to scan the surface of the lattice. A set of scan coils move the electron beam across the lattice in a two-dimensional grid fashion. The electron beam strikes the lattice, producing secondary electrons from it, which are collected by a backscatter detector, converted to a voltage, and amplified. The amplified voltage is applied to the grid of an analog display, causing the intensity of the spot of light to change. The SEM image as shown in Fig. 3.2b consists of large number of continuous spots of varying intensity on the face of the display that correspond to the topography of the sample. The deflection pattern of the electron beam is identical to the deflection pattern of the light spots on the screen. A completely bright patch is observed in the area corresponding to the tuPOY lattice, indicating that whatever electrons fall on the lattice are conducted on its surface.

Both TEM and SEM in conjunction prove that tuPOY exhibits properties similar to the conducting materials. In addition, the photographic film from TEM analysis shows a dark shadow corresponding to the copper grid as noticed on the left corner and top of the image in Fig. 3.1a. This establishes the fact that the tuPOY lattice

---

[2]SEM analysis is performed on Field Emission Gun-Scanning Electron Microscope (JSM-7600F) under a high vacuum of $9.6e - 5$ Pa.

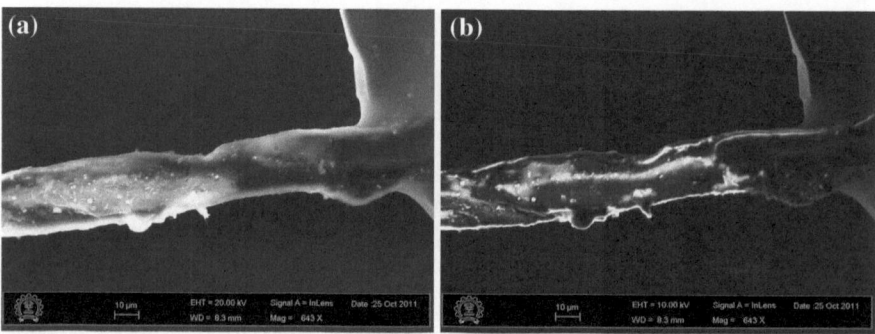

Fig. 3.2 SEM analysis of tuPOY and its comparison with regular POY. The scanning electron microscopy shows the bright lattice in Fig. 3.2a which indicates conduction properties of tuPOY and its thermally unstable state, which is absent in darker lattice of POY in Fig. 3.2b

has electronic properties similar to copper grid. tuPOY lattice in its arrested state demonstrates the conducting properties similar to metallic elements.

With this claim of tuPOY behaving as a metal, it remains to be explored whether tuPOY also retains the properties of textiles This is validated with micrographical analysis in the form of X-ray diffraction.

3.2 Wide Angle X-Ray Diffraction

X-ray diffraction[3] peaks result from constructive interference of monochromatic beams of X-rays scattered at specific angles from each set of lattice planes. The Cu K-alpha (1.5405 Å) radiation operates at 40 KV/40 mA and the scattering intensities are recorded in the range of $2\theta = 0$–120°, where 2θ is the Bragg Angle. Micrographical and XRD analysis[4] of tuPOY is performed and the results are compared with that of regular POY [2–5]. The results are substantiated in Fig. 3.3 wherein the XRD pattern of the regular fabric is shown in blue, while that of tuPOY is shown in red. In either lattice a single-dominant peak is observed confirming their similar morphologies. This further relates to the fact that the tuPOY is amorphous in nature with Bragg angle equal to 21.54°. The similar morphologies exhibited by the XRD of tuPOY and POY illustrate that tuPOY retains all the properties of partially oriented yarns and hence annotates to the textile family. However, there is an increase in the peak height of the tuPOY lattice from 1300 to 2750 counts. The increase is attributed to the movement of a large number of electrons in the orbitals which are in a state of motion. This internal excitement produces an inherent stress in the lattice, referred to as the microstress. XRD analysis validates the capture of tuPOY in an

---

[3] X-ray diffraction patterns were measured using a Philips PW1050 X-ray diffractometer.
[4] Phillips X'Pert software package is used to calculate the sample composition from X-ray diffraction data.

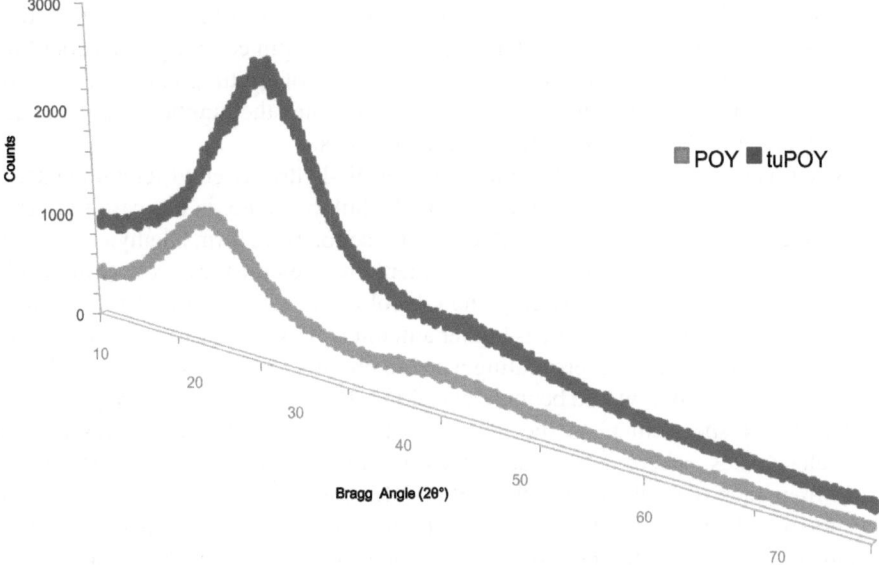

Fig. 3.3 XRD analysis of tuPOY and its comparison with regular POY. XRD analysis indicates a difference of 2500 counts signifying thermally unstable state of tuPOY

unstable molecular state, wherein there is increase in the peak counts and hence, a development of inherent microstress in the lattice. Subsequently this microstress is quantized in tuPOY using Nuclear Magnetic Resonance.

3.3 Nuclear Magnetic Resonance

Nuclear magnetic resonance spectrometer quantifies the absorption of energy by free hydrogen nuclei, when exposed to magnetic bombardment. tuPOY is placed inside a glass tube between the poles of a powerful magnet. The test process involves variation in strength of the magnetic field while maintaining a constant frequency of RF radiation. The energy required to flip the proton equals the energy of the RF radiation at some explicit value of the field. tuPOY[5] lattice when bombarded with nuclear magnetic energy breaks into parts by dislocation of hydrogen atoms. ^1H is the nucleus most sensitive to H-NMR signal. The precession of the hydrogen nuclei spin in the magnetic field is the interaction used in NMR. Due to their nuclear spin, the H protons have a magnetic field associated with them. When they are placed in a magnetic field approximately half of the protons become aligned with the field, while the other half align against the field. The transition between these two states is

---

[5]The tuPOY material is characterized by ^1H-NMR recorded on a 300 MHz VARIAN mercury spectrometer, taken on a sample dissolved in acetone.

quantified in NMR. In addition, the magnetic field is also affected by the orientation of neighboring nuclei, which is referred to as the spin-spin coupling. This coupling causes splitting of the signal for H nucleus. The size of splitting is independent of the magnetic field and the number of splitting indicates the number of chemically bonded nuclei in the vicinity of the observed nucleus.

When magnetic energy strikes the lattice of bis-hydroxyl ethyl terephthalate, it will displace an hydrogen atom and free an alcohol molecule, in our case EG. This EG molecule will combine with the free EG of reaction III to form Diethylene Glycol (DEG). Further bombardment of magnetic energy will release the third EG molecule which will combine to form Triethylene Glycol (TEG). As in case of tuPOY, with the same amount of energy, there is higher amount of TEG because of the absence of the polycondensation chain preventing a stable product. Because of this unstability, to dislodge a ^1H atom, would be more easier in tuPOY as compared to regular POY.

The NMR spectrum plots the intensity of absorption of radio waves against the delta values, where 1δ value equals 1 part per million of the instruments radio frequency (i.e., one millionth of 300 MHz) equaling 300 Hz. The δ values on the horizontal axis move from lower to higher energy levels, from right to left. Figure 3.4 compares the spectroscopic findings within regular POY and tuPOY.

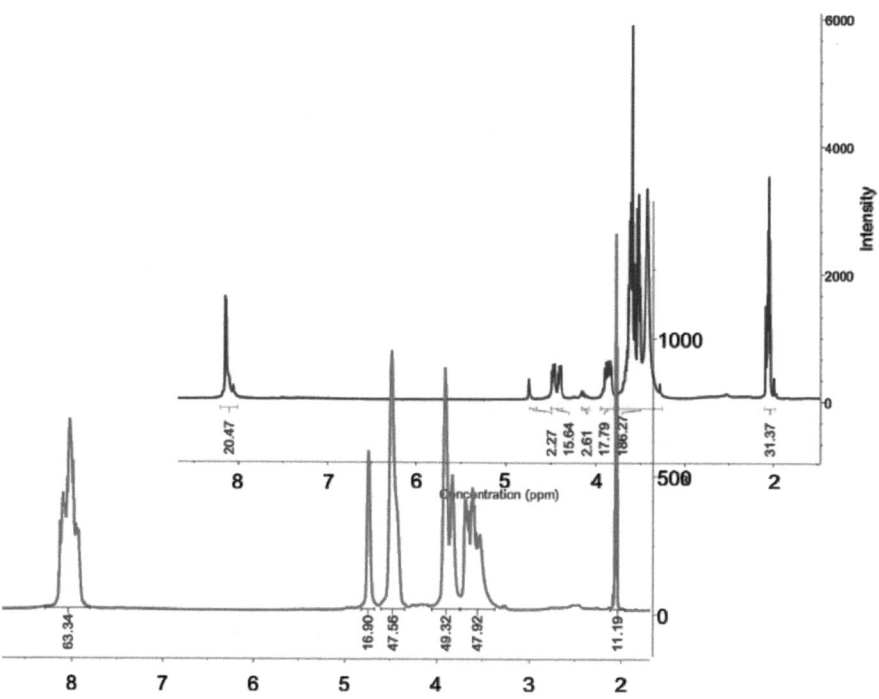

Fig. 3.4 NMR spectroscopy comparison between tuPOY and regular POY. NMR provides the relative ratio of the number of units of TEG, DEG, and EG in tuPOY relative to those observed in regular POY

Table 3.1 The higher ratios of TEG presence in the NMR spectrum of tuPOY validates the thermal unstability, as tuPOY splits more TEG molecules compared to regular POY with the same amount of energy

Compounds	Units in POY	Units in tuPOY	Relative ratio
TEG	97.42	186.27	1.912
DEG	47.56	20.4	0.4289

This is further substantiated by DEG ratios which is complementary, due to higher molecules of DEG in regular POY which are abstained from conversion

Propagating from right to the left figure illustrates the intensities i.e., the number of units of TEG, DEG, and EG in tuPOY relative to those observed in POY. Table 3.1 quantifies the peak counts of the three ethylene glycol molecules for comparison. It is observed that tuPOY has higher content of TEG (1.9 times) as compared to the regular POY. The thermal unstable property of tuPOY causes more H atoms to be displaced and hence a higher ratio of TEG molecules in comparison to regular POY. In contrast, a higher content of DEG is observed in regular POY, as there is lesser displacement of H atoms. This occurs because there is a higher release of ethylene glycol in tuPOY when it is subjected to magnetic energy.

Having quantified the thermal unstable property of tuPOY, its metallic properties are further explored by outlining its radiating behavior with infrared spectroscopy.

3.4 Fourier Transform Infra Red Analysis

The infrared spectrum indicates the fingerprint of a sample with peaks corresponding to the frequency of vibrations between the bonds of the atoms making up the lattice. Every different material has a unique combination of atoms and hence, no two materials produce the exact same infrared spectrum. This spectral fingerprint is readily distinguished from the absorption or transmittance patterns of the constituent compounds. During FTIR,[6] the infrared source emits a broad band of successively increasing wavelength of infrared radiation. The IR source is produced by electrically heating a Globar silicon carbide rod to 1500 °C. The IR radiation passes through an aperture, which controls the amount of energy and passes it to the interferometer, which further performs an optical inverse Fourier transform. The beam enters the sample compartment where it is transmitted through or reflected off its surface. The intensities of the IR beam are detected by a liquid nitrogen cooled mercury cadmium telluride detector. The detected signal is Fourier transformed to get the IR spectrum of the tuPOY sample. The morphology of tuPOY and regular POY as observed in their IR spectrum in Fig. 3.5 is the same, concluding that both the materials hold similar properties. Thus the FTIR also reconfirms that the developed material is a

---

[6]Molecular fingerprinting of tuPOY is performed using NICOLET IR200 FTIR system where infrared beam is made to fall on the material.

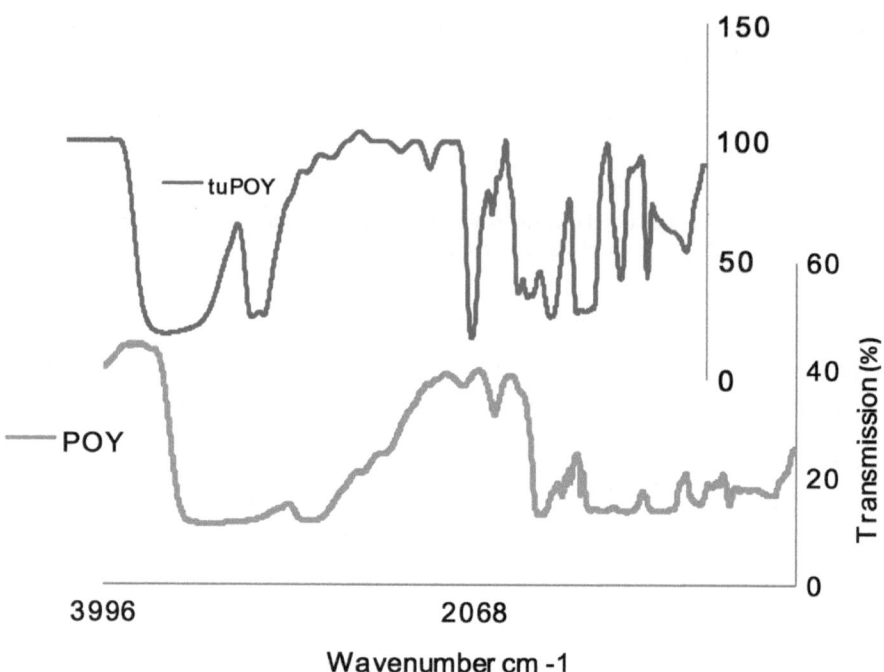

Fig. 3.5 FTIR spectroscopy compares the morphologies of tuPOY and regular POY. The transmittance in excess of 100 % verifies the radiating properties of tuPOY

textile. Figure 3.5 compares the IR transmittance intensities between the regular POY and tuPOY.

Atoms in a molecule do not remain stationary but pulsate back and forth about their interatomic distance. These atomic vibrations are classified into two types;

1. stretching vibrations, which cause a change in the length which connects them and
2. bending vibrations which constitutes a change in the bond angle.

The molecular vibrations bring about a change in the dipole moment of the molecule and hence give rise to absorption and transmission bands in the infrared spectrum. When infrared beam falls on tuPOY, the lattice does not allow the beam to completely pass through it. This occurs because the molecular arrangement in the lattice is not aligned in the direction of the beam. The infrared beam utilizes some of its energy to align the molecules in lattice with the incident beam, in order for the infrared rays to pass through. The energy required to align the lattice is termed as the vibrational energy, and the phenomenon in which the lattice atoms align with the direction of the infrared beam is known as vibrational resonance. The difference between the incident energy and the vibrational energy is projected in the Fourier spectrum in terms of the transmittance.

When a material radiates electromagnetic energy in space, the phenomenon is called electrical resonance. The results of infrared spectroscopy of tuPOY reveal that at certain frequencies the transmittance exceeds 100 %. At those specific frequencies, the infrared rays cause resonance on the lattice surface. The infrared radiation possesses sufficient energy to remove electrons from atoms in materials, through which the radiation passes. tuPOY emits the excess energy, which it has absorbed en masse, at remnant frequencies during which the material has failed to achieve resonance. At these frequencies, a transmittance percentage in excess of 100 % is observed as vibrational resonance is already achieved, due to lattice alignment and also due to the occurrence of electrical resonance, validating the radiating properties of the material.

References

1. Transmission electron microscope. http://www.nobelprize.org/educational/physics/microscopes/tem/index.html. Accessed 18 Apr 2012 (05:47:33)
2. Yannas, I.V.: Massive internal fracture of an amorphous polyester. Science **166**(3902), 227–228 (1969). http://www.sciencemag.org/content/166/3902/227.abstract
3. Oppenlander, G.C.: Structure and properties of crystalline polymers. Science **159**(3821), 1311–1319 (1968). http://www.sciencemag.org/content/159/3821/1311.abstract
4. Schindler, F., Lupton, J.M., Muller, J., Feldmann, J., Scherf, U.: How single conjugated polymer molecules respond to electric fields. Nat. Mater. **5**, 141–146 (2006). http://dx.doi.org/10.1038/nmat1549
5. Lendlein, A., Jiang, H., Jünger, O., Langer, R.: Light-induced shape-memory polymers. Nature **434**, 879–882 (2005). http://dx.doi.org/10.1038/nature03496

Chapter 4
tuPOY: Mathematical and Analytical Characterization

Abstract The mathematical characterization of a concept embeds a logical element in it, giving the invention a concrete structure. The conducting and radiating properties of tuPOY are given a mathematical emblem in this chapter. The behavioral pattern of tuPOY to a stimuli and its subsequent molecular state change variation due to its conduction, and radiation properties are mathematically modeled, providing an insight into its metallic behavior. A new mathematical concept of interchange phenomenon at molecular level is proposed to support the claims. The numerical manifestations spout a gamut of operations of tuPOY in technologies of future.

Keywords Mathematical characterization · Micro stress · Interchange phenomenon · Kinetic energy

The science of materials has evolved concurrently with its mathematical understanding and quantized control of the processes and phenomenon, involved in their operation. 1882 marked the year when Thomas Alva Edison devised the first fluorescent bulb. However, it was not until 1885, when W.H. Preece, an electrician working with the British Post Office found a way to calculate the fusing current of a wire based on its diameter. This mathematical foundation, christened the Edison effect, served a utilitarian purpose for the first time. It was on this support that Edison patented the incandescent device as a voltage indicator. Not only do we remember it as 'the first patent in electronics' but it has also been called as the 'first application of thermionic current to a useful purpose' [1].

Electromagnetic theory of materials, as we comprehend today, is understood much simply by Maxwell's model, where a curl operator spatially varies the electric and magnetic fields. In 1950s, Shockley created the point contact transistor, considered to date as the holy grail for the first processing material. This seminal work as often described has been buttressed with models which govern the flow of electrons in solid state crystals. The models developed over the years have been the governing epitome of devices based on semiconductors. The quantum state of a physical system or a moment of particle in space changes with time is simply an interpretation of time-dependent Schrodinger equations. Thus for every new invention, the need arises to replicate the characteristics of the materials and their properties, with numerical representations.

© Springer India 2016   29
H.D. Mustafa et al., *tuPOY: Thermally Unstable Partially Oriented Yarns*,
Advanced Structured Materials 23, DOI 10.1007/978-81-322-2632-1_4

tuPOY having its unstability characteristics needs calibration and derivation of a relation between the incident energy on its surface and the thermodynamic movements of kinematic energy inside the crystal lattice. This chapter provides a mathematical and analytical model on the basis of how the material behaves when a wave of certain energy is incident on it. Interchange processes occur that change the thermophysical and chemical properties, and hence develop micro structural changes in tuPOY, which embed metallic characteristics in it. Numerical equations are derived to make a quantitative assessment of the thermodynamics, kinetics, and kinematics involved with the tuPOY lattice. A steady state representation is developed as an interchange phenomenon to measure these lattice kinetics.

4.1 Inherent Microstress in tuPOY

The tuPOY material is thermally unstable and has a natural tendency to return to its original stable state, which is continually opposed by the innovatively formulated retardant. This phenomenon causes a friction in tuPOY due to which an inherent microstress is developed, which is quantified in the initial stage of the calibration process. The microstress quantity initially derived is subtracted from the flux caused by the energy of the incident wave.

The maximum kinetic energy of the lattice structure [2] is defined by (4.1) as

$$(E_R)_{max} = \frac{4A}{(1+A)^2} E_n, \tag{4.1}$$

where A is the molecular mass of the POY lattice and E_n is the kinetic energy of the incident wave. The relation between [3] the molecular spacing d, Bragg angle θ and the wavelength λ is defined by (4.2) as

$$d = \frac{\lambda}{2 \sin \theta}. \tag{4.2}$$

The change in the lattice structure (Δd) i.e., change in d, due to the additional incident energy is calculated by [4], noting the change in the angle θ as in (4.3)

$$\frac{\Delta d}{\Delta \theta} = \frac{-1}{2} \lambda \csc \theta \cot \theta = -d \cot \theta \tag{4.3}$$

The relative microstrain (ϵ) and microstress (σ) are defined by [5], as (4.4) and (4.5), respectively.

$$\epsilon = \frac{\Delta d}{d} = \left(\frac{|d_{final} - d_{initial}|}{d_{initial}} \right) \tag{4.4}$$

$$\sigma = \frac{\epsilon}{2} E_R \qquad (4.5)$$

Having acquired the residual microstress, we constitute the internal kinetic energy excitement that occurs through an Interchange mechanism of state change characterization, when external energy is fed to the tuPOY.

4.2 Mathematical Science of tuPOY

An interchange phenomenon is proposed which relates to the arresting state of the tuPOY, quantifying the interchange between Unreacted Functional Units (UFU) and bonds, abbreviated as U–B and the stimulated effects between the bonds themselves abbreviated as B–B [6, 7]. It has been proved in [8] that various interchange reactions occur in polymers during its state of molecular unstability. Studies have been extensively conducted on biological interchange phenomenon like chromosome diffusion under excitation and viral infections in human body [9]. These basic naturally occurring processes are built upon to quantify the interchange between UFU and bonds and between the bonds themselves. Equal reactivity of the functional units with bonds of tuPOY and the satisfaction of the Markovian property of the chemical processes is assumed [10]. Let L_x denote the chain x-mer and R_x a ring x-mer. The U–B interchange model is defined in (4.6) and (4.7). In the UB interchange, an UFU on L_x interchanges with bonds in other molecules to form L_{x+j} as shown in (4.6).

$$L_x + L_{j+n} \rightleftharpoons L_{x+j} + L_n \qquad (4.6)$$

Further a UFU in L_{x+j} reacts with a bond in the same molecule to form R_x as shown in (4.7)

$$L_j + R_x \rightleftharpoons L_{x+j}. \qquad (4.7)$$

The B–B interchange model is as defined in (4.8)–(4.10).
A bond in L_x interchanges with a bond in the other molecules is as shown in (4.8)

$$L_{x+j} + L_{m+n} \rightleftharpoons L_{x+n} + L_{m+j} \qquad (4.8)$$

There is also a possibility that a bond in L_{x+j} interchanges with another bond in the same molecule to form R_x as shown in (4.9). This interchange is in similitude with U–B interchange, however, due to presence of bonded elements, is still unstable.

$$L_j + R_x \rightleftharpoons L_{x+j} \qquad (4.9)$$

Further a bond on R_x interchanges with a bond in the other molecule forming R_j. Also a bond R_{x+j} interchanges with another bond in the same ring, to from R_x and R_j as in (4.10).

$$R_x + R_j \rightleftharpoons R_{x+j} \qquad (4.10)$$

Since the molecular lattice of tuPOY is unstable, there are p interchanges between a polymer chain $N_{x,k}$ and ring $N_{R_x,k}$, when excited with external energy. Furthermore, there will be variations when tuPOY returns to its initial arrested state denoted without superscript p. The state equations, which define the number of such interchanges at any state k due to incident external energy, are shown in (4.11) and (4.12) for chains and rings, respectively [9, 11].

$$N_{x,k} = N_{x,0} + \sum_{i=1}^{k} (\delta N_x + \delta N_x^p)_i \tag{4.11}$$

$$N_{R_x,k} = N_{R_x,k} + \sum_{i=1}^{k} (\delta N_x + \delta N_x^p)_i, \tag{4.12}$$

where $N_{x,k}$ and $N_{R_x,k}$ denote the number of AB type x-mers and ring-mers, with increments of δN_x and δN_{R_x}, respectively at any instant k. With the aid of the interchange phenomenon, the state change mechanism of tuPOY is analyzed by taking the permutations and combinations of all possible ring and chain formation into consideration. When an additional external energy is fed to lattice of tuPOY, it changes from a state $i \rightarrow (i+1)$. With system volume at state i, the transition probabilities for the U–B interchanges are defined as $P\{x^U, L\}$ and $P\{(x+j)^U, R_x\}$, for probability that an UFU on L_x interchanges with bonds in other molecules, and for probability that an UFU in L_{x+j} reacts with one bond in the same molecule to form R_x, respectively. Similarly the probabilities of B−B interchange are defined as $P\{x, L\}$, $P\{x+j, R_x\}$, $P\{R_x, L\}$ and $P\{R_{x+j}, R_x\}$, without the 'U' superscript indication. The number of polymer rings resulting due to this interchange is given by (4.13)–(4.15),

$$L_{x+j}^U \rightleftharpoons R_x + L_j$$
$$\delta N_{R_x}^1 = \sum_{j=1} P_i\{(x+j)^U, R_x\}, \tag{4.13}$$

where $P_i\{(x+j)^U, R_x\}$ denotes the probability that an unreacted functional unit on L_{x+j} reacts with one bond in the same molecule to form a ring.

$$L_{x+j} \rightleftharpoons R_x + L_j$$
$$\delta N_{R_x}^2 = \sum_{j=1} P_i\{(x+j), R_x\} \tag{4.14}$$

where $P_i\{(x+j), R_x\}$ indicates the probability that one bond in L_{x+j} interchanges with another bond in the same molecule to form a ring.

$$R_{x+j} \rightarrow R_x + R_j$$

$$\delta N_{R_x}^3 = \sum_{j=1}^{x} (x+j) P_i \left\{ R_{x+j}, R_x \right\} \tag{4.15}$$

$P_i \left\{ R_{x+j}, R_x \right\}$, represents the probability that one bond in R_{x+j} interchanges with another bond in the same ring to form different rings. To form the rings, there are x ways when $j = x$ and $(x+j)$ ways when $j \neq x$. Further to avoid counting same combinations twice a factor of 1/2 is introduced as shown in (4.16).

$$R_{x-j} + R_j \rightarrow R_x$$

$$\delta N_{R_x}^4 = \frac{1}{2} \sum_{j=1}^{x} P_i \left\{ R_j, L \right\} (x-j) \frac{N_{R_{x,i}}}{h_i} \tag{4.16}$$

$P_i \left\{ R_j, L \right\}$ is the probability that one bond in R_x interchanges with bonds in the other molecules. The number of reacted functional units in state i is indicated by h_i. When the energy is radiated from the lattice, the return of tuPOY to its arrested state is as shown in (4.17)–(4.20) for the various permutations.

$$L_{x+j}^U + R_x \rightarrow L_{x+j}$$

$$\delta N_{R_x}^5 = -\sum_{j=1}^{x} P_i \left\{ R_{x+j}, R_x \right\} \frac{N_{R_{x,i}}}{h_i} \tag{4.17}$$

$$R_x + L_j \rightarrow L_{x+j}$$

$$\delta N_{R_x}^6 = -P_i \left\{ R_x, L \right\} \sum_{j=1}^{x} (j-1) \frac{N_{R_{j,i}}}{h_i} \tag{4.18}$$

If $j \neq x$, an odd number of rings and if $j = x$, an even number of rings are broken. Furthermore, x combinations of rings are released if $j = x - j$ and $x/2$ otherwise, to attain the initial unstable state i as shown in (4.19) and (4.20).

$$R_j + R_x \rightarrow R_{x+j}$$

$$\delta N_{R_x}^7 = -P_i \left\{ R_x, L \right\} \sum_{j=1}^{x} (j-1) \frac{N_{R_{j,i}}}{h_i} \tag{4.19}$$

$$R_j + R_x \rightarrow R_{x+j}$$

$$\delta N_{R_x}^8 = \frac{1}{2} \sum_{j=1}^{x} x P_i \left\{ R_x, R_j \right\} \tag{4.20}$$

Collecting the interchanges occurring in the molecular lattice from (4.13) to (4.20), the total number rings formed during transition from state i to $(i+1)$ and back are obtained by summation as shown in (4.21).

$$\delta N^p_{R_x} = \sum_{m=1}^{8} \delta N^m_{R_x} \tag{4.21}$$

Similarly the polymer chains $N_{x,k}$ formed and deformed during the transesterification process are as shown in (4.22)–(4.35),

$$L^U_{x-j} + L_{j+m} \rightarrow L_x + L_m$$

$$\delta N^1_x = \sum_{j=1}^{x} \sum_{m=1}^{x} P_i \left\{ (x-j)^U, L \right\} \frac{N_{R_{j+m,i}}}{h_i} \tag{4.22}$$

where $P_i \left\{ (x-j)^U, L \right\}$ gives the probability that an unreacted functional unit on L_x interchanges with bonds in another molecules.

$$L^U_j + L_{m+x} \rightarrow L_{j+m} + L_x$$

$$\delta N^2_x = \sum_{j=1}^{x} \sum_{m=1}^{x} P_i \left\{ j^U, L \right\} \frac{N_{R_{x+m,i}}}{h_i} \tag{4.23}$$

$$L^U_{x-j} + R_j \rightarrow L_x$$

$$\delta N^3_x = \sum_{m=1}^{x} P_i \left\{ (x-j)^U, L \right\} \frac{N_{R_{j,i}}}{h_i} \tag{4.24}$$

$$L^U_{x+j} \rightarrow L_x + R_j$$

$$\delta N^4_x = \sum_{j=1}^{x} P_i \left\{ (x+j)^U, R_j \right\} \frac{N_{R_{j,i}}}{h_i} \tag{4.25}$$

$$L_{x-j+m} + L_{n+j} \rightarrow L_x + L_{m+n}$$

$$\delta N^5_x = \sum_{j=1}^{x} \sum_{m=1}^{x} \sum_{n=1}^{x} P_i \left\{ (x-j+m), L \right\} \frac{N_{j+i,i}}{h_i} \tag{4.26}$$

$$L_{x-j} + R_j \rightarrow L_x$$

$$\delta N^6_x = \sum_{j=1}^{x} (x-j) P_i \left\{ (x-j), L \right\} \frac{N_{R_{j,i}}}{h_i} \tag{4.27}$$

$$L_{x+j} \rightarrow L_x + R_j$$

$$\delta N^7_x = \sum_{j=1}^{x} x P_i \left\{ (x+j), R_j \right\} \tag{4.28}$$

$$L_j^U + L_x \rightarrow L_{j+m} + L_{x-m}$$

$$\delta N_x^8 = -(x-1)N_{x,i}\sum_{j=1}^{x}\frac{P_i\{j^U,L\}}{h_i} \tag{4.29}$$

$$L_x^U + R_j \rightarrow L_{x+j}$$

$$\delta N_x^9 = -P_i\{x^U,L\}\sum_{j=1}^{x}\frac{jN_{j,i}}{h_i} \tag{4.30}$$

$$L_x^U + L_j \rightarrow L_{x+m} + L_{j-m}$$

$$\delta N_x^{10} = -P_i\{x^U,L\}\sum_{j=1}^{x}\frac{jN_{j,i}}{h_i} \tag{4.31}$$

$$L_x^U \rightarrow L_{x-j} + R_j$$

$$\delta N_x^{11} = -\sum_{j=1}^{x}P_i\{x^U,R_j\} \tag{4.32}$$

$$L_x + L_j \rightarrow L_m + L_n$$

$$\delta N_x^{12} = -(x-1)P_i\{x,L\}\sum_{j=1}^{x}\frac{jN_{j,i}}{h_i} \tag{4.33}$$

$$L_x + R_j \rightarrow L_{x+j}$$

$$\delta N_x^{13} = -(x-1)P_i\{x,L\}\sum_{j=1}^{x}\frac{jN_{R_j,i}}{h_i} \tag{4.34}$$

$$L_x \rightarrow R_j + L_{x-j}$$

$$\delta N_x^{14} = \sum_{j=1}^{x-2}P_i\{x,R_j\} \tag{4.35}$$

The total number of polymer chains formed during the interchange process can thus be calculated as in (4.36)

$$\delta N_x^p = \sum_{m=1}^{14}\delta N_x^m \tag{4.36}$$

The internal kinetic energy Δ of the tuPOY is measured as the total number of interchanges dm in infinitesimal time interval Δdt. An equal reactivity of UFU with bonds of tuPOY is assumed for satisfaction of the Markovian property for the chemical processes. Taking V as the sensor patch volume, the transition probabilities are written in kinetic forms. The probabilities affected due to U–B interchange are represented in (4.37) and (4.38), whereas those affected by B–B interchanges are characterized in (4.39)–(4.42),

$$P_i\{x^U, L\} \approx \frac{\left\{ \frac{\kappa_L^U N_{x,i} h_i}{V} \right\}}{\Delta} \tag{4.37}$$

$$P_i\{(x+j)^U, R_x\} \approx \frac{\left\{ \kappa_{R_x}^U N_{x+j,i} \right\}}{\Delta} \tag{4.38}$$

$$P_i\{x, L\} \approx \frac{\left\{ \frac{\kappa_L N_{x,i} h_i}{V} \right\}}{\Delta} \tag{4.39}$$

$$P_i\{R_x, L\} \approx \frac{\left\{ \frac{\kappa_L x N_{R_{x,i}} h_i}{V} \right\}}{\Delta} \tag{4.40}$$

$$P_i\{x+j, R_x\} \approx \frac{\left\{ \frac{\kappa_{R_x} N_{x+j,i}}{V} \right\}}{\Delta} \tag{4.41}$$

$$P_i\{R_{x+j}, R_x\} \approx \frac{\left\{ \frac{\kappa_{R_j} \kappa_{R_{x-j}} N_{R_{x+j,i}}}{\kappa_{R_x}} \right\}}{\Delta}, \tag{4.42}$$

where κ denotes the interchange rate constants (i.e., thermal conductivity) for the various combinations as defined earlier, and N denotes the number of chains in the polymer. Equating the sum of the above probabilities to unity, we obtain an expression for internal kinetic energy Δ of tuPOY.

$$\Delta = \frac{\kappa^U (N_x - h)h}{V} + \kappa^U N_{R_x} + \frac{\kappa_L N_x^2}{V} + \frac{\left(\frac{1}{2}\right) \kappa_L h^2}{V} + \frac{\kappa_b N_{R_x} h}{V} + \kappa_{R_x} N_x$$
$$+ \frac{\left(\frac{1}{2}\right) \kappa_{R_j} \kappa_{R_{x+j}} N_{R_x}}{\kappa_{R_x}} \tag{4.43}$$

When external energy is incident on the material, it produces an internal kinetic energy excitement (Δ). The effective change in internal kinetic energy of the sensor is calculated by subtracting the residual microstress developed in the tuPOY from the change in the internal kinetic energy of the sensor as denoted in (4.44).

$$\Delta_{eff} = |\Delta - \sigma| \tag{4.44}$$

Recording this change in internal kinetic energy on the lattice for externally incident energy or signal evolves the science of tuPOY, enabling its application in sensing, radiating, and processing applications as investigated in the subsequent chapters.

References

1. Dylla, H.F., Corneliussen, S.T.: John Ambrose Fleming and the beginning of electronics. J. Vac. Sci. Technol. A: Vac. Surf. Films **23**(4), 1244–1251, (2005). http://pubs.acs.org/doi/abs/10.1021/j100202a078
2. Jenkin, A.: Irradiated polymers. Nature **243**(5404), 247–248 (1973)
3. Bragg, W.H.: Bakerian lecture: X-rays and crystal structure. Philos. Trans. R. Soc. Lond. Ser. A Contain. Papers Math. Phys. Character **215**, 253–274 (1915). http://www.jstor.org/stable/91108
4. Mallick, B., Patel, T., Behera, R., Sarangi, S., Sahu, S., Choudhury, R.: Microstrain analysis of proton irradiated pet microfiber. Nucl. Instrum. Methods Phys. Res. Sect. B: Beam Interact. Mater. Atoms **248**(2), 305–310 (2006). http://www.sciencedirect.com/science/article/pii/S0168583X06006410
5. Hutchinson, J.M., Mccrum, N.G.: Microstress mechanism for the time dependence of the modulus of crystalline polymers following an imposed change in volume. Nature **252**(5481), 295–296 (1974)
6. Kahn, O., Martinez, C.J.: Spin transition polymers: from molecular materials toward memory devices. Science **279**(5347), 44–48 (1998)
7. Garnier, F., Hajlaoui, R., Yassar, A., Srivastava, P.: All-polymer field-effect transistor realized by printing techniques. Science **265**(5179), 1684–1686 (1994). http://www.sciencemag.org/content/265/5179/1684.abstract
8. Suematsu, K., Okamoto, T.: Theory of ring formation in a reversible system: general solutions. Colloid Polym. Sci. **270**, 405–420 (1992). doi:10.1007/BF00665983
9. Suematsu, K., Okamoto, T.: Interchange reactions during polymerization: re-examination of the polymerization mechanism of ϵ-caprolactam. J. Phys. Chem. **96**(23), 9498–9507 (1992)
10. Suematsu, K., Okamoto, T.: Theory of ring formation in irreversible system: AB model. Colloid Polym. Sci. **270**, 421–430 (1992). doi:10.1007/BF00665984
11. Suematsu, K., Okamoto, T., Kohno, M., Kawazoe, Y.: Functionality dependence of the distribution of cyclic species in network formation. J. Chem. Soc. Faraday Trans. **89**, 4181–4184 (1993). http://dx.doi.org/10.1039/FT9938904181

Chapter 5
tuPOY as a Radiating Element: Antenna

Abstract The ability of tuPOY to emit electromagnetic radiations provides a playground to envision its role in antenna and wave propagation. This chapter analyzes the radiating properties of tuPOY and its use an antenna. A prototype of a nonmetallic antenna with tuPOY as its integral part of design is developed. The novel construction of the antenna structure highlights the fact that apart from ministering its inherent objectives, tuPOY antenna emanates as a standalone power source. The proposed structure is capable of not only being self-sustaining from immanently generated power but also providing residual power for networked devices as well.

Keywords Antenna · Radiation pattern

Since their development, construction of an antenna has been synonymous with metallic elements, creating a major deterrent of their use in multitude of applications [1]. Researchers, over the years, have been unable to subjugate a fully nonmetallic material for antenna design. Current trends of nonmetallic antennas involve doping of metallic formulations or threads in weaving conditions, in order to obtain conducting fibers. They are further combined with small electronic components, electrochemical, optical sensors, and circuitry to produce the so called smart metal-based materials. Several attempts to integrate micro metallic yarns into textile structures have been widely experimented. Textile materials such as felt, songket, fleece, lint, satin, spacer, and even denim have been employed as antenna substrates [2, 3]. Behind this facade lies a radiating patch and/or the ground plane fabricated with conductive threads, interwoven with textile materials or substrates in metallic castings. Polyester forms a substrate for antennas in military applications with Nora fabric as the radiating material [4]. A composite of three metals, nickel, copper, and silver, Nora does not qualify as a textile material [5, 6]. The aforesaid elastic fabrics when stretched cause a change in the permittivity and subsequent changes in the resonant frequencies. Nano antenna arrays are basically deposition of conducting materials such as gold on graphene substrates yielding CNT antennas operating in the THz frequency range [7]. Majority of the textile antennas presented in literature are masqueraded as metallic

© Springer India 2016   39
H.D. Mustafa et al., *tuPOY: Thermally Unstable Partially Oriented Yarns*,
Advanced Structured Materials 23, DOI 10.1007/978-81-322-2632-1_5

layers [8]. Furthermore, operationally these antennas yield pattern fragmentation, power absorption, central frequency shift, transmission errors, and loss of link margin [1, 8–10]. This chapter describes a completely nonmetallic antenna, developed from tuPOY and synthetic composite textiles like raw silk and nylon without any metallic camouflage or doping.

5.1 Antenna Design

Chapter 3 in this book is discusses in detail the spectroscopic techniques to determine the nature, characteristics, lattice, and structural transformations of the fabricated tuPOY. Infrared spectroscopic measurements using FTIR measure infrared light absorbance due to molecular vibrations in tuPOY at different frequencies of infrared light. The resultant absorption spectrum represents its molecular fingerprint of tuPOY. Thus infrared Fourier spectroscopy verifies that tuPOY belongs to the family of textiles and also asserts its radiating properties. FTIR illustrates that at certain frequencies energetic particles are radiated into space, i.e., the electrons manage to be displaced from the atoms in the tuPOY lattice, through which the infrared beam passes. The empirical results are juxtaposed with theoretical claims substantiated by the interchange phenomenon. The transition from unreacted to reacted state ($i \rightarrow i + 1$) refers to the process where UFU turn into reacted functional units. In the process of transition of the atoms in the lattice, energy is released. This energy is continuously emitted in a spectrum of electromagnetic radiation until equilibrium is reached.

This work outlines an exemplary microstrip polymer textile antenna, created from tuPOY and synthetic textiles composites of raw silk and polynylon. The structure of tuPOY antenna is the amalgamation of ground plane and a radiating substrate. The twin layer radiating patch substrate has tuPOY as its top layer over a viscous layer of raw silk. A similar patch of viscous raw silk forms the uppermost layer of the three-layered ground plane. tuPOY forms the middle layer while the latter end of the ground substrate is prepared out of polynylon which acts as an insulator. The antenna ground plane is formulated with 34 % tuPOY, 33 % raw silk and 33 % polynylon. Molten tuPOY and polynylon is embedded on either side of the raw silk cloth. The dielectric ground plane substrate so formed has an ϵ_r of 3.4 with a thickness (h) of 2 mm and a loss tangent ($\tan \delta$) of 0.02.

The structure of the proposed textile antennas is as shown in Fig. 5.1. The radiating patch of the antenna is made in a proprietary shape of nonsymmetrical alphabet 'M,' with a material composition of 49 % raw silk and 51 % tuPOY, equally distributed in alternate knitting pattern. The patch substrate has an ϵ_r of 3.09 with a thickness (h) of 0.2 mm and a $\tan \delta$ of 0.014. The feed point probe of thickness 0.4 mm to the antenna is placed at coordinates of (9.75 and 9.45),[1] which provides for an even

---

[1]Referenced to the origin at point (0,0) on the ground plane.

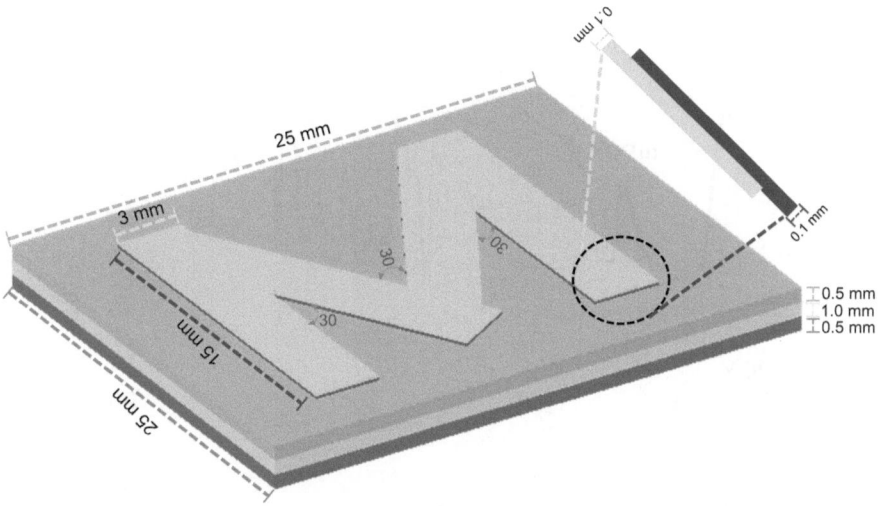

Fig. 5.1 Schematic of the tuPOY and synthetic fabric composite asymmetrical 'M'-shaped antenna. The substrate is fused to the patch with raw silk and hence no gluing will be required resulting in minimization of losses. 'M' shape is designed to protect proprietary interests of the inventors. Additionally this design yields dual band operations due to symmetry

power distribution across the patch. The feed and the probe are made of special conductive wires of tuPOY of 4 mm thickness.

With this innovatively formulated design, the required patch size is 20 mm × 25 mm for operation in test frequency range of 4–10 GHz. This size is smaller in comparison to the metallic configurations, let alone the existing so called metallic textile patches. Alternating, the patch and substrate thickness and the feed point, the textile antennas can be adjusted to work in frequency range and bandwidth of desired choice as the application demands.

5.2 Microstrip tuPOY Antenna Operation

FTIR helps in identifying the molecular vibrations in tuPOY lattice. The antenna patch on the upper end, composed of tuPOY, is directly fed with electromagnetic energy through the feeding probe. This feed causes an excitation in the uppermost tuPOY substrate layer resulting in resonance on its lattice and subsequent electromagnetic radiation. The tuPOY molecules moving forward and backward develop inherent stretching and bending vibrations. The alternating electric field generated due to the incident electromagnetic energy interacts with the fluctuations in the dipole moment of tuPOY molecule. When the radiation matches with the vibration

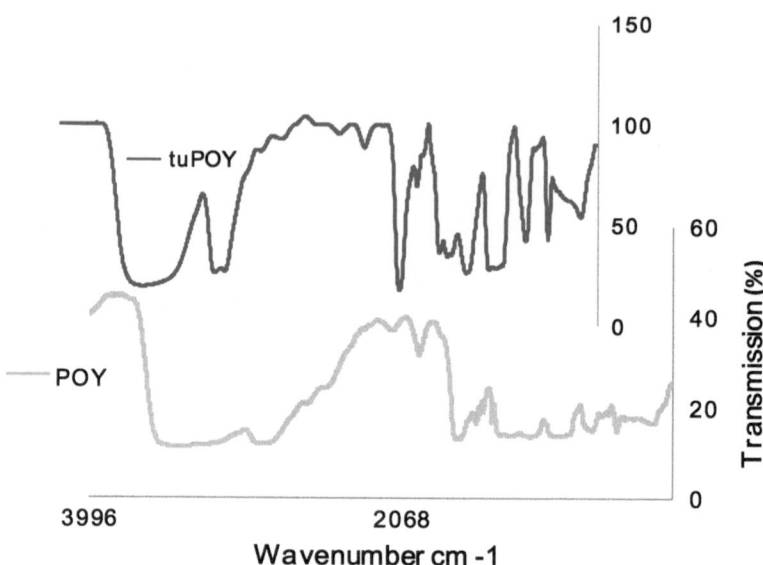

Fig. 5.2 FTIR spectroscopy compares the morphologies of tuPOY and regular POY. The transmittance in excess of 100 % verifies the radiating properties of tuPOY

frequency of the molecule then the electromagnetic field will be absorbed, causing a molecular resonance in the lattice. This provides sufficient energy to dislodge the electrons form the atoms to cause electromagnetic radiations. FTIR characteristics shown in Fig. 5.2 reflect that frequencies corresponding to transmittance rate exceeding 100 % have already achieved electrical resonance and the resultant additional molecular resonance causes the EM radiation.

The radiation pattern is omnidirectional with two lobes originating from the feed point. The upper lobe radiates in a wireless environment, while that in the backward direction strikes the surface of raw silk which acts as an absorber for this incoming energy. The energy absorbed on the raw silk layer converts to thermal energy and is transported to the tuPOY layer in the ground plane. The tuPOY layer in the ground plane is larger in size compared to that of the patch and hence the thermal energy received here does not result in resonance. This tuPOY layer acts as a conductor for the thermal energy. This additional thermal energy is transported to a power generation unit where it contributes to energy requirements of the system. Furthermore, this thermal energy is prevented from leaking from the tuPOY by a layer of polynylon, which not only acts as an absorbent of radiations, but also provides mechanical stability to the antenna. This morphology removes the need for adding a radiation shield, reflector and power source in our design, yielding a first of its kind antenna.

(a)

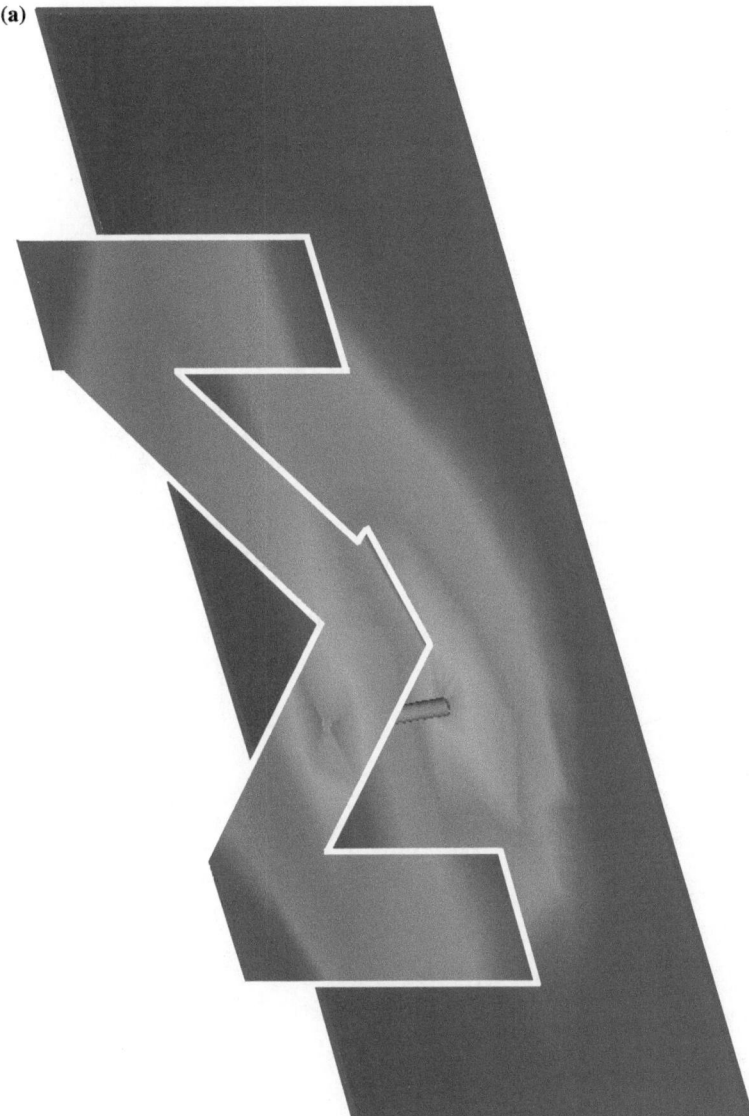

Fig. 5.3 (**a**) An even distribution of the electromagnetic field over the tuPOY patch with minimum leakage. (**b**) An impressive VSWR of around 1.016 for the frequency of 7.9 GHz. (**c**) The S_{11} characteristics of the tuPOY antenna are below the 10 dB scale in the 7.5–9.5 GHz range. (**d**) An omnidirectional and uniform radiation pattern with no minor lobes for the operational frequency range. (**e**) An omnidirectional pattern with a uniform distribution over entire 180° on the decibel scale. (**f**) Perfect impedance matching of the tuPOY antenna at 7.9 GHz. The lowest VSWR is obtained at the same frequency, correlating a 50 Ω impedance matching. (**g**) Radiation Efficiency (RE), Antenna Efficiency (AE), Conjugated Matching Efficiency (CME), and the Voltage Source Efficiency (VSE) with the highest efficiency of radiation attained at 60 %

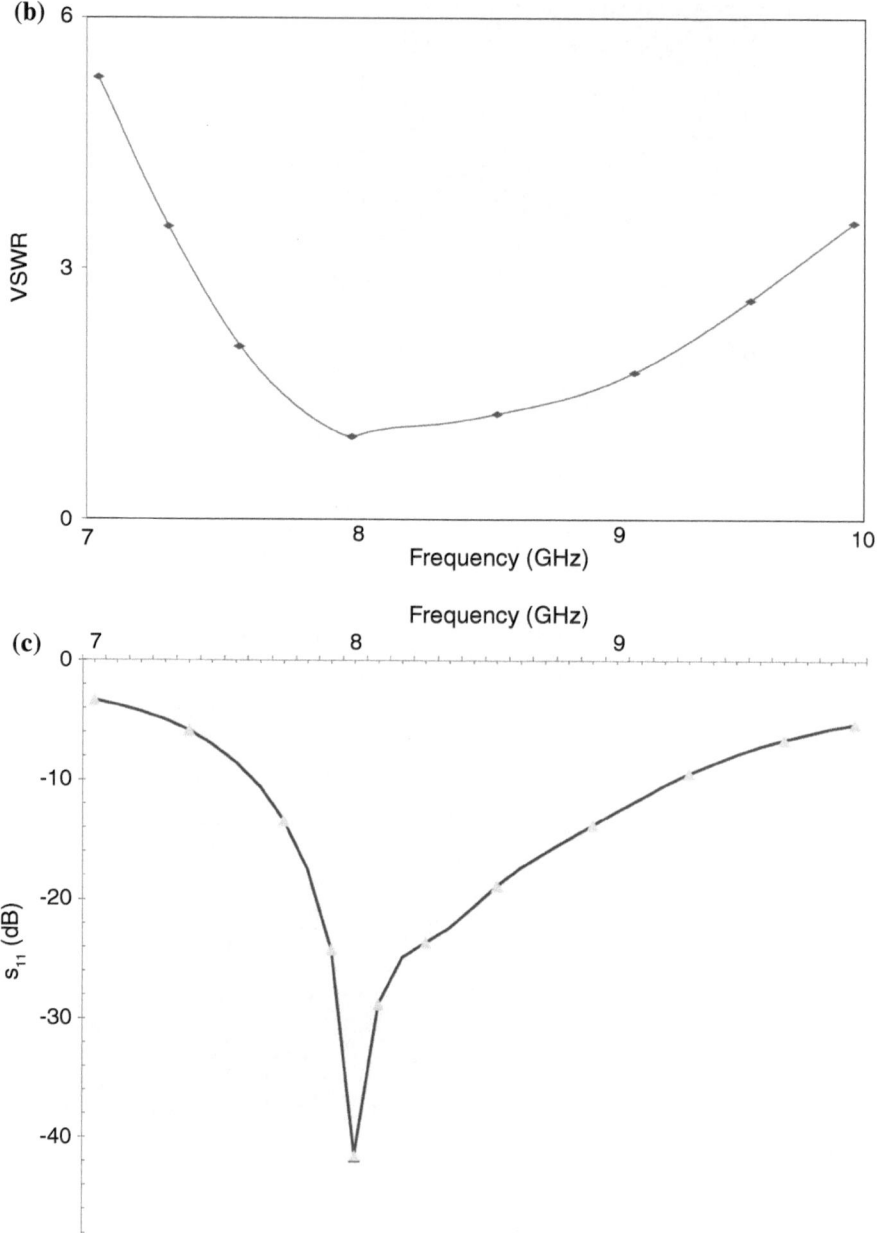

Fig. 5.3 (continued)

(d)

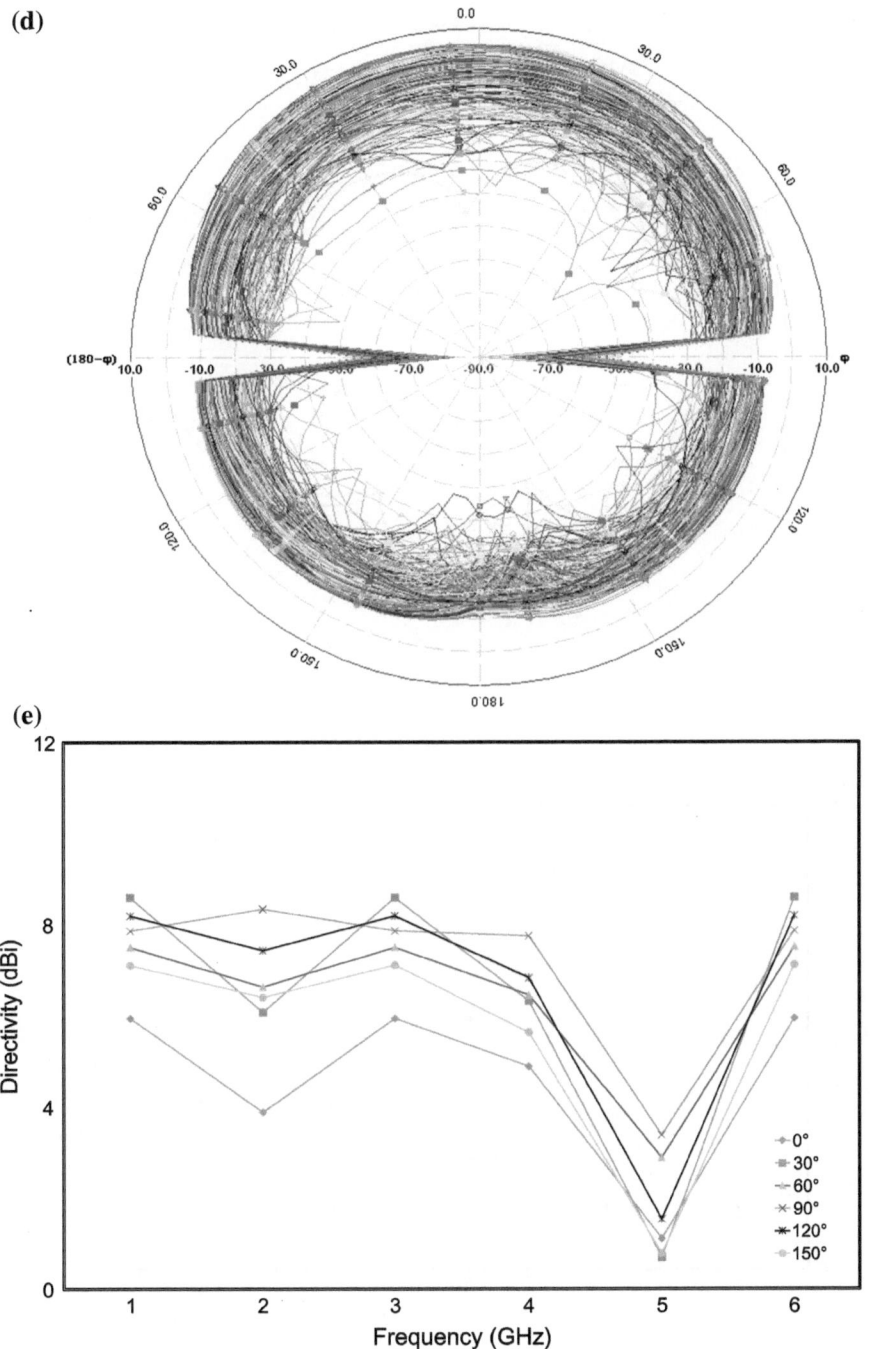

(e)

Fig. 5.3 (continued)

(f)

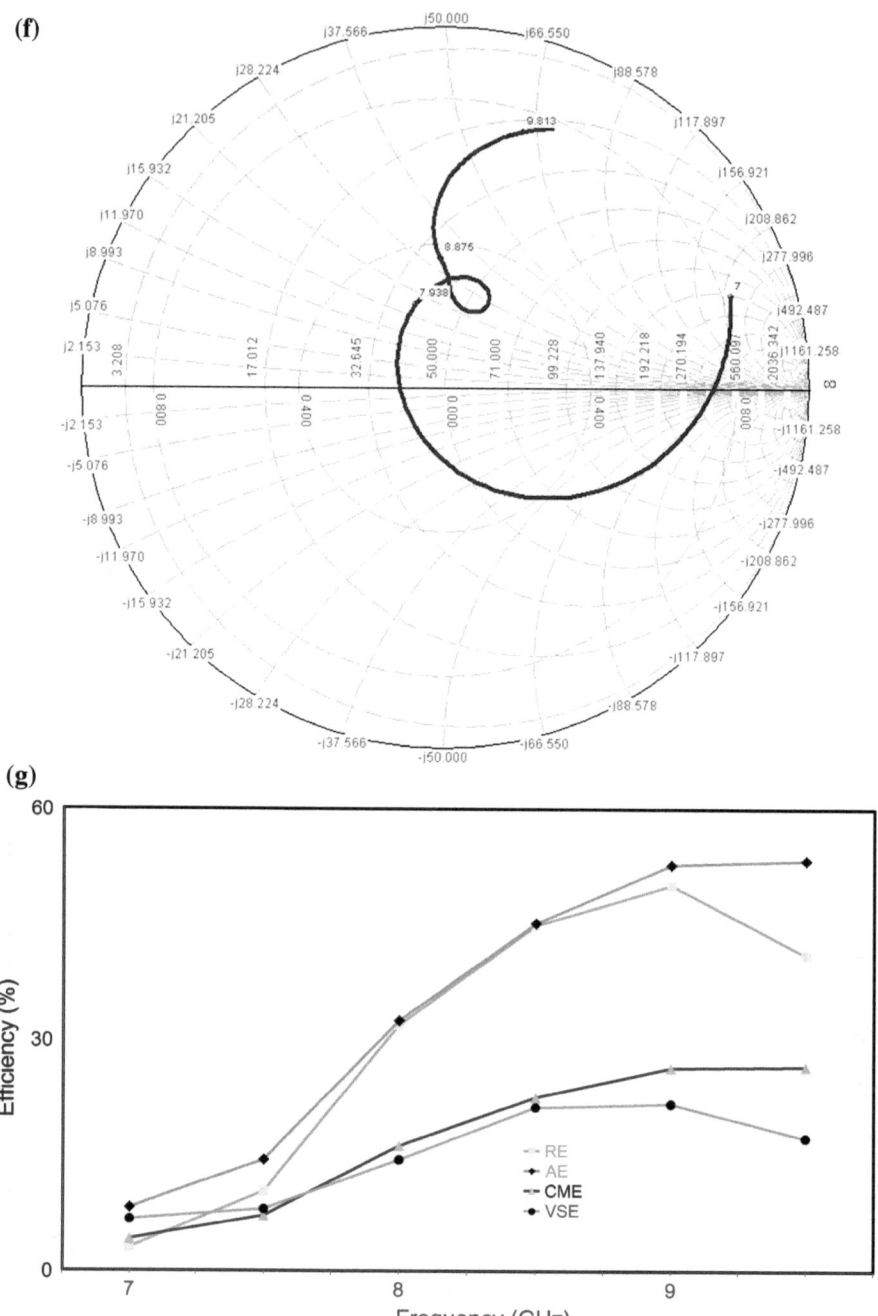

(g)

Fig. 5.3 (continued)

5.3 Operating Performance Characteristics of tuPOY Antenna

Measurements on the proposed antenna are performed in the range of 7.5–9.5 GHz on a vector network analyzer[2] and also verified by simulation results on a method of moments-based simulator.[3] The results of the experimentation and simulations presented in Fig. 5.3. A shows an impressive VSWR of around 1.016 for the center frequency of 7.9 GHz. B shows an omnidirectional and uniform radiation pattern with no minor lobes for the operational frequency range which is complimented in C with a uniform distribution over entire 180° on the decibel scale. This is followed by D which shows a perfect impedance matching of the tuPOY antenna at 7.9 GHz. The lowest VSWR is obtained at the same frequency, correlating a 50 Ω impedance matching. E confirms that the s_{11} characteristics of the tuPOY antenna are below the 10 dB scale in the desired Ultra Wide Band (UWB) range. F shows the Radiation Efficiency (RE), Antenna Efficiency (AE), Conjugated Matching Efficiency (CME) and the Voltage Source Efficiency (VSE) with the highest efficiency of radiation attained at 55 %. The even distribution of the electromagnetic field over the tuPOY patch with minimum leakage is observed in G.

Table 5.1 Comparison of existing and proposed antennas

Slotted 'M'	Substrate 30 × 30 × 0.813	Frequency 5.47–5.68	Bandwidth 210	Fractional 3.8	Gain 6.32
		6.15–6.23	80	2	7.46
Inverted 'E'	4 (mm) thick	2.2	100	4.5	–
		3	110		
Asymmetrical 'V'	80 × 30 × 4.5	2.49	80	3.2	–
		2.4		2	
'E'	140 × 200 × 10	1.8–2.46	600	30	6.7
'E'	Slotted Substrate	5.05–5.88	830	15	7.5
'E'	35 × 54 × 0.8	4.9–5.94	1000	19	7
'E'	5880	9.6–13.3	3700	31	7
	DUROID				12
'V'	70 × 5 × 6	2.95	–	3.5	10.5
		4.721		27	

---

[2] 'M'-shaped antenna is analyzed over Agilent 8722ET (50 MHz–40 GHz) transmission/reflection network analyzer.

[3] The method of moments simulator used is commercial version of IE3D. More than 50 simulations are performed to optimize the feed point and obtain a VSWR of 1.016.

The proposed 'M'-shaped antenna is compared and contrasted for its performance metrics against the 'E' and 'V'-shaped antennas and the results are enumerated in Table 5.1. The results clearly demonstrate the superiority in performance of tuPOY-based antennas as compared to the existing ones. The size of the ground plane of the antennas proposed in [11] is 8 times and the thickness is 10 times as compared to the proposed 'M'-shaped antennas. The bigger size makes the positioning of the antenna very difficult and hence makes it practically unusable in applications of pervasive computing. Similarly the antenna design proposed in [12] has three times thicker substrate as compared to the proposed antenna, for obtaining the same operational bandwidth. Furthermore its morphology is highly unsuitable for use in practical applications. The 'E'-shaped antennas as proposed in [13] are highly inefficient in their operation as compared to the 68 % efficiency of the proposed design. They obtain a good bandwidth and a comparable gain, which is good for use in high speed wireless networks but fail to perform in applications requiring low data rate like transmission of physiological signals etc. The antenna as proposed by [14] is useful in pervasive computing especially body area networks, but due to their very low operational bandwidth is not practically feasible. 'E'-shaped antennas made with Duroid as the substrate material give a large bandwidth and good directivity [15]. However, these antennas have efficiency of one third as compared to the tuPOY-based 'M'-shaped antenna at 28.84.

This chapter proposes a completely metal gratis antenna designed and developed with tuPOY as the heart of the substrate. The antenna operating in the UWB range is light weight, low cost, rugged, and hence can be ambidextrously positioned in any pervasive computing architecture. The innovatively designed tuPOY antenna can be developed for varied applications and technologies of the future. Being manufactured from a synthetic fiber it resists variation due to stretch, shrinking, and wrinkles. These properties enable us to camouflage the tuPOY antenna in daily wear. Its nonmetallic nature ensures that it cannot be detected by X-Ray machines, metal detectors making it impossible to perceive. tuPOY can be drawn into yarns of desirable length and color, thus amalgamating with regular apparels.

References

1. Haga, N., Saito, K., Takahashi, M., Ito, K.: Characteristics of cavity slot antenna for body-area networks. IEEE Trans. Antennas Propag. **57**(4), 837–843 (2009)
2. Hertleer, C., Tronquo, A., Rogier, H., Langenhove Van, L.: The use of textile materials to design wearable microstrip patch antennas. Text. Res. J. **78**(8), 651–658 (2008). http://trj.sagepub.com/content/78/8/651.abstract
3. Rahman Osman, M.A., Bin Rahim, M.K.: Wearable textile antenna: fabrics investigation. J. Commun. Comput. **7**(7), 75–80 (2010)
4. Roh, J.S., Chi, Y.S., Kang, T.J.: Wearable textile antennas. Int. J. Fash. Des. Technol. Educ. **3**(3), 135–153 (2010). http://www.tandfonline.com/doi/abs/10.1080/17543266.2010.521194
5. Liu, N., Lu, Y., Qiu, S., Li, P.: Electromagnetic properties of electro-textiles for wearable antennas applications. Front. Electr. Electron. Eng. China **6**, 563–566 (2011). doi:10.1007/s11460-011-0182-7

6. Kennedy, T., Fink, P., Chu, A., Champagne, N., Lin, G., Khayat, M.: Body-worn e-textile antennas: the good, the low-mass, and the conformal. IEEE Trans. Antennas Propag. **57**(4), 910–918 (2009)

7. Song, L., Ci, L., Gao, W., Ajayan, P.M.: Transfer printing of graphene using gold film. ACS Nano **3**(6), 1353–1356 (2009), pMID: 19438194. http://pubs.acs.org/doi/abs/10.1021/nn9003082

8. Klemm, M., Troester, G.: Textile UWB antennas for wireless body area networks. IEEE Trans. Antennas Propag. **54**(11), 3192–3197 (2006)

9. Declercq, F., Rogier, H.: Active integrated wearable textile antenna with optimized noise characteristics. IEEE Trans. Antennas Propag. **58**(9), 3050–3054 (2010)

10. Sanz Izquierdo, B., Batchelor, J.C., Sobhy, M.I.: Button antenna on textiles for wireless local area network on body applications. Microw. Antennas Propag. IET **4**(11), 1980–1987 (2010)

11. Dunne, L., Brady, S., Smyth, B., Diamond, D.: Initial development and testing of a novel foam-based pressure sensor for wearable sensing. J. NeuroEng. Rehabil. **2**(1), 4 (2005). http://www.jneuroengrehab.com/content/2/1/4

12. Locher, I., Klemm, M., Kirstein, T., Troster, G.: Design and characterization of purely textile patch antennas. IEEE Trans. Adv. Packag. **29**(4), 777–788 (2006)

13. Farringdon, J., Moore, A., Tilbury, N., Church, J., Biemond, P.: Wearable sensor badge and sensor jacket for context awareness. In: The Third International Symposium on Wearable Computers. Digest of Papers, pp. 107–113 (1999)

14. Salonen, P., Yang, F., Rahmat Samii, Y., Kivikoski, M.: WEBGA—wearable electromagnetic band-gap antenna. In: Antennas and Propagation Society International Symposium, vol. 1, pp. 451–454. IEEE (2004)

15. Tronquo, A., Rogier, H., Hertleer, C., Van Langenhove, L.: Robust planar textile antenna for wireless body lans operating in 2.45 GHz ISM band. Electron. Lett. **42**(3), 142–143 (2006)

Chapter 6
tuPOY in Future of Computing: Internet of Things and Ubiquitous Sensing

Abstract The utilitarian proficiency of tuPOY is exploited in an environment of internet of things in this chapter. Use of tuPOY in pervasive environments is presented as a case study. Devices made of tuPOY, seamlessly integrated in the ubiquitous environment, are detailed with explicitly designed experimental test beds in this chapter. These devices are shown to compute, communicate, and have embedded intelligence in them. The chapter illustrates the functionality of tuPOY as a bridge between the physical and digital worlds. These applications advocate and give an insight into the logical processing capabilities of the innovative material. tuPOY as a diode, a transistor, an amplifier, and in general the heart of the next generation processor takes the invention to its eventual conclusion and lays down it future endeavors.

Keywords Ubiquitous and pervasive tuPOY applications · Power generation module · tuPOY experimental test bed

This chapter presents a new paradigm from the perspective of ubiquitous on-body computing through the innovative polymer textile, tuPOY. An effective pervasive computing architecture [1–4] allows operation on low power consumption, efficient, and noncomplex communication with back-end, low data loss, or interference, and most importantly, practical and low cost hardware implementation. To achieve these characteristics, the existing modern infrastructure for wireless communications is integrated seamlessly in a pervasive environment in our proposed model with tuPOY.

The main design parameters that are emphasized include, new nonmetallic textile material tuPOY, for processing, miniaturizing the size of individual modules, optimization of the system operation at lowest power, noninclusion of wires or metallic components and providing autonomy to individual systems while ensuring unobtrusive connectivity for heterogeneous signals and applications.

© Springer India 2016

H.D. Mustafa et al., *tuPOY: Thermally Unstable Partially Oriented Yarns*,
Advanced Structured Materials 23, DOI 10.1007/978-81-322-2632-1_6

51

6.1   Ubiquitous and Pervasive Application Design

tuPOY is integrated in a pervasive environment as a wireless body area network (WBAN) architecture. The proposed WBAN from a healthcare monitoring perspective is only a case study and can be seamlessly extended to any arena of technology [5, 6]. The ubiquitous nature of WBAN is designed to provide clinicians the ability to access services and resources incessantly and irrespective to their location. The individual modules are developed with the aim of ease of use, comfort in daily activities and with a capacity to integrate a substantial number of individual modules. Equipped with these parameters of good design under consideration, the process flow of the architecture is as shown in Fig. 6.1.

From a healthcare monitoring scenario, the waveform selected for testing on the architecture is the radial artery. It has been classified as one of the most significant arteries for noninvasive diagnosis, for identification of the presence and location of disorders [7]. Radial pulse wave commonly aids in the diagnosis of diseases such as hypertension, arteriosclerosis, coronary heart disease, thyroid function abnormality, anemia and hence forms an appealing option for our test base [8].

The thermal variations or pulses from the radial artery are measured via tuPOY sensors, which are integrated with the clothing of the subject. The interchange

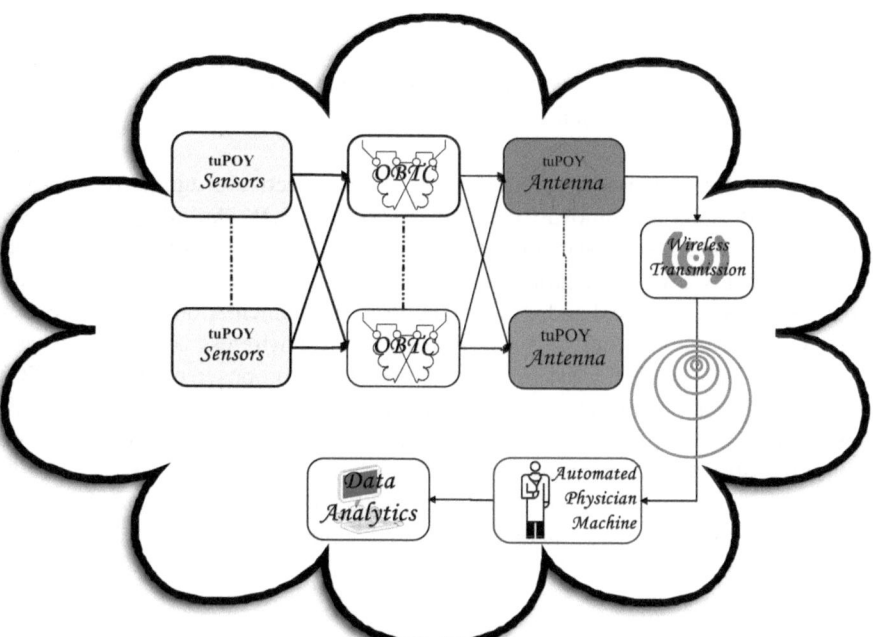

Fig. 6.1 Process flow of the pervasive WBAN architecture showing transmission of the incident energy from sensing to radiation using tuPOY and automated back-end data analytics

phenomenon helps in the mathematical correlation of the emitted radiations. The sensors transmit the signal through conductive fibers made of tuPOY, acting as conducting wires, to an on-body transmission circuit (OBTC). The OBTC transforms the thermal energy to electrical signals in UWB operating frequency range of 3.1–10.6 GHz and transmits it to an on-body textile antenna made out of a unique composite of synthetic textiles raw silk, polynylon, and tuPOY. The power generation unit of OBTC is designed to power the entire system, by absorbing heat emitted by the subject's body and from the thermal energy received via tuPOY. The connections between the OBTC and antennas are made up of tuPOY fibers, as they dually act as a transmitting medium for both electrical and thermal signals. The lack of any metallic components in the designed system prevents backscatter and retro reflections. The material is tested with experiments performed in UWB range, with operating Voltage Standing Wave Ratio (VSWR) threshold of less than 1.5.

The tuPOY antennas send or receive signals through a transmission $(T_x - R_x)$ device. A stand alone $T_x - R_x$ device that communicates over a wireless channel is presented. However, for this purpose regular devices like mobile phones can also be used [9, 10]. The transmitted signals are received at the remote health monitoring station. A rigorous classification of this data and inferencing is performed using an APM, which provides for an automatic diagnostic system, without round the clock presence of a physician or trained personnel.

The structure of the proposed WBAN has all the components connected to the subject's body, made entirely of tuPOY embedded in the clothing of the subject. This allows for better usability and practicality of the paradigm at the users end. In development of our proposed WBAN architecture, due concentration is placed on developing unique modules at each stage, based on strong theoretical foundation, offering a robust architecture. Compliance is adhered with the IEEE 802.15.6 protocol and standards for WBAN, while designing the architecture [11–13].

6.1.1 Sensing Mechanism

The radial artery, classified as one of the most significant arteries for noninvasive diagnosis is selected for the experimental test bed of the WBAN architecture. The incident pulse wave is characterized by 6.1,

$$p = p_o + \sum_n p_n e^{i(nwt-\alpha)} \tag{6.1}$$

where $p_n = \sqrt{r_n^2 + s_n^2}$ corresponds to the kinetic energy of the incident wave and $\alpha_n = \arctan(\frac{s_n}{r_n})$ represents the phase angle of the wave with r_n and s_n as real and imaginary components of the wave [14].

According to the oriental medical practitioners, the clinician performs a holistic measurement by palpitating the area above the radial artery at the wrist location,

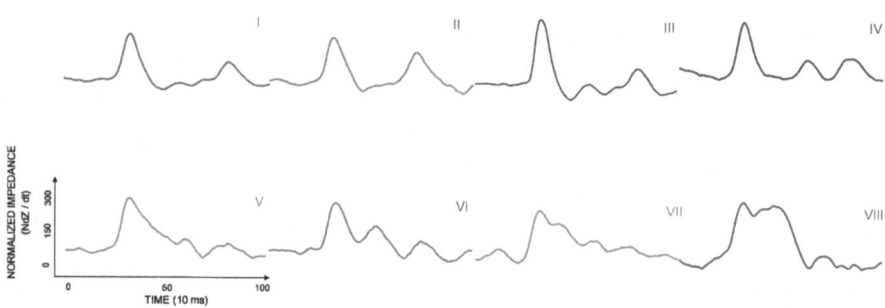

Fig. 6.2 Morphological patterns observed using the tuPOY sensor from the radial pulse

and monitors the rhythm, pulse pressure, pulse propagation, etc. for arriving at the diagnosis. Each palpation position reflects the health condition and functioning of different organs [15] of the body. At different applied pressures, different amplitudes or energies are sensed which are then correlated with the body conditions [16]. Tartiere et al. [17] exploited the radial pulse wave non-invasively, for the evaluation of left ventricular systolic performance in heart failure. The traditional method is emulated as the ground truth where a peripheral pulse analyzer represents the fingers as pulse pressure sensors. The ground truth is used as a base to verify the information obtained from the tuPOY-based WBAN.

6.1.1.1 Radial Pulse Analysis

The analysis of the waveforms from the radial pulse, reveal existence of eight different patterns unique among diseased and normal subjects, as shown in Fig. 6.2. Pragmatically different diseases tend to bring about inherent changes in the morphology of the radial pulse. The morphologies corresponding to aforementioned patterns are discussed in Table 6.1.

The morphological variations strongly influence indication of a disease according to the pattern morphology. Clinical analysis[1] reveal that patterns I and IV are recorded in control subjects without any signs of specific disease. Pattern VII is a precursor to pattern VIII, and both are predominant in subjects with Myocardial Infarction (MI). Thus, the predominant pattern represents the health status of the individual and the intermittent appearance of the interspersed pattern specifies the tendency of the individual for acquiring the respective disease. The interspersed patterns could be dormant, and are an indication of the onset of morbidity in an individual. A time series data with predominant patterns I and IV with interpositions of pattern II, indicate the onset of morbidity. Interpositions of pattern III indicate the spread of the disease

---

[1]Clinical correlation is performed in normal and diseased subjects at Bhabha Atomic Research Center in presence of a physician.

Table 6.1 Morphological characteristics of the radial pulse waveforms

Pattern	Morphology
I	Ideal Plethysmographic waveform
	The first peak is ending below the base line
	Height of secondary peak is around 50 % of the first peak
II	The first peak is ending below the base line
	Height of second (multiple) peak is around 75 % of the first peak
III	Similar to the ideal Plethysmographic waveform
	Narrower primary peak
	The height of secondary peak is around 20 % of the first peak
IV	All peaks are ending at the base line
V	First peak does not reach the base line
	Aberration in the downward slope near the bottom side
VI	Small aberration in the downward slope much near to the base line
VII	All the peaks do not touch base line
VIII	Aberration in the downward slope of the primary peak in the center

specifically in the liver, while pattern V and VI indicate diseases of lungs; TB and bronchial asthma respectively. Patterns VII and VIII are detected in subjects with arterial blocks and MI.

6.1.1.2 Sensor Calibration for Radial Pulse

An important criterion is to calibrate the tuPOY sensor for proper analysis of the received signal. When the arterial pulse wave strikes the sensor, the energy of the wave produces an internal thermal excitement in the lattice, which is measured using the proposed interchange phenomenon as Δ. Corresponding calibration of the sensor is performed by a peripheral pulse simulator, for identification of various types of pulse waveforms in normal and diseased conditions as shown in Table 6.2. The range of internal kinetic energy excitement (Δ) is calculated and experimentally the lowest and the highest values of Δ obtained are 200–450 J. The residual microstress (σ) for the relative range of Δ, is calculated to be 25–50 J. Thus the tuPOY sensor is calibrated to receive the thermal kinetic energy signals in range of 175–425 J, for the radial artery (Table 6.3).

Table 6.2 Sensor calibration

Thermal kinetic energy (Δ_{eff})	Pattern observed
310–340	I
390–425	II
230–275	III
215–230	IV
275–295	V
295–310	VI
340–390	VII
175–215	VIII

Table 6.3 Effective thermal kinetic energy of tuPOY

Internal kinetic energy excitement (Δ)	Residual microstress (σ)	Thermal kinetic energy (Δ_{eff})
200–450 J	25–50 J	175–425 J

6.1.2 Signal Transmission Module

The internal kinetic energy excitement on the tuPOY sensor lattice is transmitted with fibers of 0.5 mm thickness made of the same material. These fibers uphold all the properties of synthetic polyester yarns and hence can be crimped, soldered, dyed, and subjected to all kinds of textile processing. tuPOY excels in properties of high tensile strength, good impact, and abrasion resistance with a maximum diameter of 0.5 mm, making it an excellent choice for capturing the heat variations. It operates in a temperature range from −65 to 200 K, giving it a perfect compatibility for wearable technology. It gives an additional advantage of conduction of both thermal and electrical waveforms, thus preventing the use of metallic wires at any stage. The characteristic impedance z_{of} for this fiber is given by (6.2),

$$z_{of} = \frac{138}{\sqrt{e}} \log_{10} \frac{D}{d} \tag{6.2}$$

with dielectric constant e, dielectric diameter D and center conductor diameter d [18, 19]. The attenuation a_f offered by the fiber (db/100) ft is calculated as in (6.3),

$$a_f = \frac{0.434}{z_{of}} \sqrt{F} \left(\frac{\sqrt{R_1}}{d} + \frac{\sqrt{R_2}}{D} \right) + 2.78F\sqrt{e}P_f \tag{6.3}$$

where R_1 and R_2 are ratio of center and outer conductivity to copper with dielectric power factor P_f. tuPOY fiber resistance can be varied from a minimum value of 100

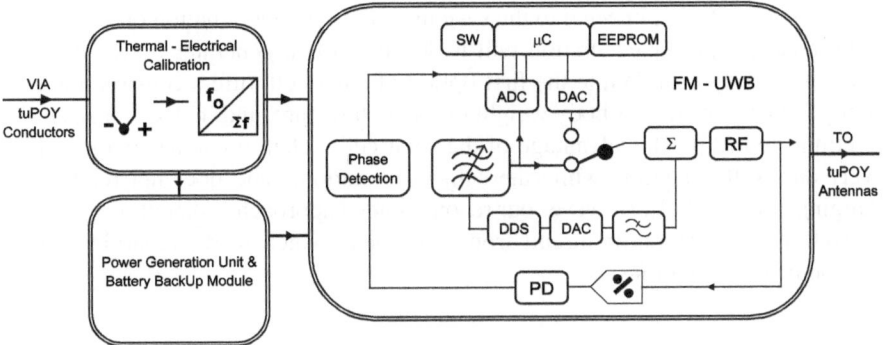

Fig. 6.3 Process flow and instrumentation of the on-body transmission circuit (OBTC). The thermal to electrical calibration is followed by the FM-UWB modulation

to more than 500,000 ohms per thousand feet. The velocity of propagation V_p is given by (6.4) as a percentage of free wave velocity.

$$V_p = \frac{100}{\sqrt{e}} \tag{6.4}$$

The thermal variations from the sensors are received at the OBTC, which processes them into its corresponding electrical waveform. This thermal energy is also used for powering the system, in addition to heat generated from the subject's body.

For propagating the signal using the existing wireless medium, a dual frequency modulation (FM) technique is adapted [20]. The integrated FM-UWB using dual FM is shown in Fig. 6.3. The advantage of using this technology is that it provides comparatively lower power consumption than commonly used impulse radio [21].

In the first stage, the signal is modulated using a low frequency sub carrier using narrow band frequency shift keying (FSK). The performance of the FM-UWB is characterized by the modulation index of this subcarrier. This intermediate signal, generated at 2.5 MHz FM is used to modulate an RF oscillator. The analog FM modulation has a modulation index of about 4, which generates a signal in the UWB frequency range attuned for transmission with the help of the textile tuPOY antennas.

6.1.3 Power Generation for On-Body Modules

Generating power at low cost and minimum structural complexity forms an essential criterion for a practical implementation of any pervasive computing system, especially for the on-body modules [22]. In the proposed architecture, due care is taken for the lowest possible power consumption, while developing the individual modules. To satisfy the power requirements of our system, a twofold approach, firstly where a solar cell is connected to the system and installed as a button on the apparel and secondly a power generation unit is developed that generates power from the exuberated body heat. While the first option of solar cell, offers construction simplicity, it increases the on-body weight of the architecture, and in due course should be eliminated [23]. The advantage of the solar cell is that, it can be charged effectively during the daytime while the subject is outdoors and does not require any charging circuitry [24]. However our recommended approach is that of the proposed on-body power generation circuit from body heat, which works regardless of the environmental conditions.

6.1.3.1   Structure of the Power Generating Unit (PGU)

The thermal energy can be derived from two sources in our architecture viz., from the tuPOY sensors and excess body heat generated from the subject. Both these sources are used to our advantage for generation of power for the architecture. Solid-state thermoelectric generators with no moving parts, that are acoustically silent and inexpensive to manufacture are employed [25]. The schematic of the device and a combined circuit diagram is shown in Fig. 6.4. The thermocouples made of constantum at the cold junction and germanium at the hot junction, are connected in a zigzag manner to save space yielding the effective dimension of the unit as $2.5\,\text{mm} \times 2.5\,\text{mm}$ in cross-section. With such small size it can be easily implanted anywhere on the subject, without much discomfort. The materials used in the fabrication of the individual thermopiles are low cost, high conduction, nonhazardous, and have high thermal conductivity. The thermopiles are connected to each other with tuPOY patches, acting as conductors, and the entire structure is insulated with a thin insulating film. The insulation helps in prevention of thermal energy leakage from the system.

For calculating the power generated the Seebeck effect is expanded, which quantifies the generation of electric power from heat flow across a temperature gradient [18]. This heat flow drives the free charge carriers towards the low temperature end, which in turn generates a voltage (V) that is directly proportional to temperature gradient (ΔT) as $V \approx \varsigma \Delta T$, where ς is the Seebeck coefficient. The temperature at the cold end is denoted as T_c and that of the hot end as T_h. A current $I = JA$ and heat power Q enters the hot end, generating a power P at the cold end. The generation efficiency η is given by [26] in (6.5),

$$\eta = \frac{P}{Q} = \frac{J \int_{T_c}^{T_h} \varsigma \, dT - J^2 \int_0^l \rho dx}{J T_h \varsigma_h + \kappa_h \bigtriangledown T_h} \tag{6.5}$$

where κ is the thermal conductivity of tuPOY and l is the length of the thermocouple unit. We further note that the maximum efficiency is defined by Carnot efficiency

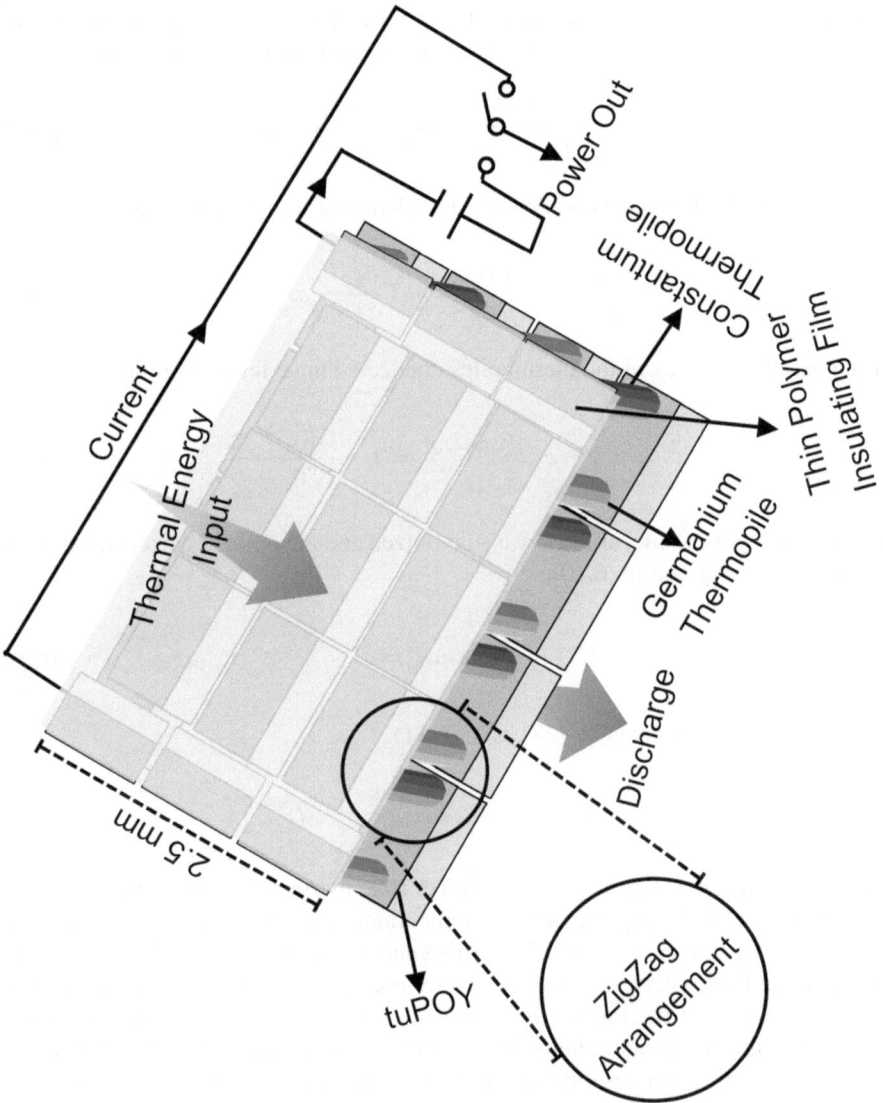

Fig. 6.4 Schematic and circuit diagram of power generation module. The zigzag arrangement between the hot and cold junction thermopiles, help in reducing the size. The thermal and electrical energy are dually conducted by tuPOY fibers

$\eta_c = \frac{T_c - T_h}{T_h}$ and the corresponding reduced efficiency is given by $\eta_r = \frac{\eta}{\eta_c}$. As the limit $\Delta t \to 0$ the reduced efficiency of η is given by (6.6)

$$\eta_r = \frac{\mu \left(\frac{\varsigma}{z_f}\right)\left(1 - \frac{\mu\varsigma}{z_f}\right)}{\frac{\mu\varsigma}{z_f} + \frac{1}{z_f T}} \qquad (6.6)$$

where $z_f = \frac{\varsigma^2}{\kappa\rho}$ is the thermoelectric figure of merit and $\mu = \frac{J}{\kappa\nabla T}$ is the relative current density. The quantity κ [27] is constrained by heat and given as (6.7),

$$\frac{d(\kappa \nabla T)}{dx} = -T\frac{d\varsigma}{dT}J \nabla T - J^2\rho \tag{6.7}$$

where $T\frac{d\varsigma}{dx}$ is the Thomson coefficient. The relation of μ varying only with T is given in (6.8).

$$\frac{d(\frac{1}{\mu})}{dT} = -\frac{1}{\mu}\frac{d\mu}{dT} = -T\frac{d\varsigma}{dT} - \mu J^2\rho\kappa \tag{6.8}$$

The overall efficiency of the thermopiles connected in series is given in (6.9).

$$\eta = 1 - e^{-\int_{T_c}^{T_h} \frac{\eta r(\mu,T)}{T}dT} = 1 - \frac{\varsigma_c T_c + \frac{1}{\mu_c}}{\varsigma_h T_h + \frac{1}{\mu_h}} \tag{6.9}$$

By optimizing the value, efficiency is maximized and the optimal current density can be calculated as given in (6.10).

$$\int_{T_c}^{T_h} \kappa\mu dT = Jl \tag{6.10}$$

and the voltage V is given by [28] as in (6.11).

$$V = \frac{1}{\mu_h} + \varsigma_h T_h - \frac{1}{\mu_c} - \varsigma_c T_c \tag{6.11}$$

In the schematic of 2.5 mm × 2.5 mm lattice, 1024 thermopiles are cascaded, delivering about 2 millivolts of power across a temperature difference of 1.5 K and a thermal to electric efficiency of about 0.2 %. This generated power is sufficient to continuously power the on-body modules of the proposed pervasive computing architecture.

Having outlined the sensing mechanism, a transmission module for the sensed emission and an efficient stand-alone power generation system, the architecture demands the need for an efficient on-body transmitting fully nonmetallic textile antenna, which is described in the subsequent section.

6.1.4   Textile Antenna

Antennas and propagation for body centric wireless communications form an essential element of any pervasive computing architecture [29]. The main points of consideration in the design of the antenna for on-body stationing are first, low mutual influence between antennas and the human body for high antenna efficiency and low specific absorption rate (SAR), second, small size and low profile and finally, polarization of the antenna only in the direction required to avoid unnecessary radiations toward the body surface [30]. The novel textile antenna effectively addresses these issues. These antennas are made completely from tuPOY composites and to the best of our knowledge is an archetype in the published literature.

The wireless technology of UWB is adapted because of low data rate of radial pulse signal. Additionally, UWB offers benefits due to low power operation and extremely low radiated power [31]. Combining the UWB transmission with purely textile microstrip patch antennas, makes it a very effective and user-friendly choice for the WBAN architecture. The proposed textile antennas are only 2 mm thick, can be dyed in any color, and hence can be seamlessly integrated into the regular apparels. The combination of the above architecture with UWB transmission in frequency range of 7.5–9.5 GHz yields a low power consuming system. The frequency range of 7.5–9.5 GHz is selected due to non requirement of large bandwidth, and because it is the safest range for WBAN in the IEEE 802.15.6 standards. It is non-interfering with the existing bluetooth, *wifi* and other wireless standards. Owing to the placement of these antennas and the ground plane touching the body, the thermal energy on the surface of tuPOY might leak due to its high affinity to attract heat. The insulating layer of polynylon avoids this phenomenon from occurring, thus avoiding any thermal or electromagnetic radiation directed toward the body during the operation of antenna. The transmission of the signal from the UWB antenna to the on-body receiving device and the issues concerning the same is discussed in the next section.

6.1.5 UWB Signal Modulation

An on-off modulation scheme is adapted, which presents the best option from perspectives of energy efficiency implementation and on-body temperature raise [13]. The signal input distribution of the transmitting symbol is denoted as (6.12).

$$\mu_k = \begin{cases} 1, & \text{with probability } \omega \\ 0, & \text{with probability } 1 - \omega \end{cases} \tag{6.12}$$

An M-ary Pulse Position Modulation (PPM) scheme is used such that $\omega = 1/M$, where M is the number of time slots in a symbol period T, and a pulse is transmitted per symbol [10]. The transmitted signal [32] can hence be represented as (6.13),

$$x(t) = \sum_M p \left(t - a_m \left(\frac{T}{M} \right) - mT \right) \tag{6.13}$$

where $a_m \in A = \{0, 1, \ldots, M - 1\}$ is the PPM symbol. The pulse waveform $p(t)$ has a finite support in $[0, T_p)$. In UWB transmission, triggering a gated oscillator by a finite duration baseband waveform generates short pulses [21]. A triangular waveform given in (6.14), is employed,

$$x_b(t) = \begin{cases} 1 - \left| \frac{2t}{T_p} - 1 \right|, & 0 \le t \le T_p \\ 0, & \text{otherwise} \end{cases} \qquad (6.14)$$

where T_p is the triggering function or pulse duration. The pulse waveform is given by (6.15),

$$p(t) = x_b(t) \cos(2\pi f_c^n t) \qquad (6.15)$$

where f_c^n is the frequency of the nth subband. Hence, the transmitted signal $x(t)$ is given by (6.16),

$$x(t) = \sum_M \sum_{n=0}^{N_{cpb}-1} p(t - nT_c - a_m T_{BPM} - mT_{sym}) \qquad (6.16)$$

where N_{cpb} are consecutive pulses with duration $T_{burst} = N_{cpb} T_c$. Further, T_c is the inverse of the Peak Reference Frequency (PRF) with mean $PRF = N_{cpb}/T_{sym}$ and the symbol time $T_{sym} = 2T_{BPM}$. Thus the PPM-UWB system can be represented by (6.17),

$$S_{TX}^{(m,k)}(t) = \sqrt{E_b} \sum_{i=0}^{\infty} \sum_{j=0}^{N_{bit}-1} \sum_{n=0}^{N_s-1} x_{tx} \left(\begin{array}{c} t - iT_{int}^{(k)} - jT_b^{(k)} \\ -nT_p^{(m)} - \triangle_j \tau_m - \tau^k \end{array} \right) \qquad (6.17)$$

where E_b is the bit energy, T_{int} is the time interval between successive transmissions from sensor k, T_b is the bit period for user m, $\triangle_j \in \{0, 1\}$ is the information of the bit, τ_m is the time shift for the PPM and τ^k is the asynchronous random start time of the sensor.

The burst transmission technique is adapted, in which each data packet is transmitted at a much higher rate as compared to its original sampling rate. An advantage of this method is that, since all sensors are transmitting data asynchronously without the prior knowledge of the channel, the shorter transmission time will reduce the probability of collision [33]. A two level collision avoidance scheme is employed, in which the first level avoids intra-sensor collisions, by assigning a unique transmission interval for each sensor, while the second level avoids collisions with external interferences by using a Differential Pulse Repetitive Frequency (DPRF) [34].

6.1.6   Back-End Receiver Architecture

The signals transmitted from the various on-body antennas are received simultaneously at the receiver node. The adapted receiver node, is a stand alone device, that consists of a receiver antenna and a demodulator circuit [33]. This point onwards, none of the fabrication materials used in this case studies has the restriction of compatibility with the subject's apparel, and hence, we need not restrict our use to the proposed textile materials. However, if required the textile antennas as described in Sect. 5.1, can effectively be applied in receive mode in the receiving device. Henceforth, attention is shifted to the factors affecting transmission of the signal, rather than the fabrication materials.

The receiving antenna in our test bed is a simple microstrip patch antenna made of any of the commonly available substrates and operating in the frequency range of 7–10 GHz. The dimensions can be accommodated as per application and the user's preference. The existing systems like mobile phones etc., can be used as a substitute to this module, by tuning their receive mode in UWB frequency range or alternatively modulating our transmitted signal accordingly [34].

The major characteristic of design consideration is that the receiver must identify whether the signal is from one of sensors of the intended user or it is an interference signal from nearby users. This is mainly because the transmitter is unable to identify the surrounding conditions and hence is unable to change its transmission pattern or the frequency band. This forms the main reason for adapting a different pulse rate for different users. Further, a simple low power narrow band feedback is provided to the transmitting antenna, which allows for the synchronization of the sensor nodes and the transmission pattern reconfiguration to suit the environment [33].

A major factor affecting the transmission of the signal is the burst collisions, which can occur anytime during the package transmission. If such collisions occur then the packets originating from all the concerned sensors will be lost. Hence, it is important to establish a priori information on the probabilities of such collisions. Several approaches are presented in literature for establishing probability of such collisions and an upper bound on the same is obtained by [35] as in (6.18),

$$\widehat{P}_c(i) < 1 - \prod_{j \in M\{i\}} 1 - \frac{\Delta_j + TF_{ij}}{T_j} \tag{6.18}$$

where $\widehat{P}_c(i)$ is the probability that at least one user $j \in M\{i\}$ collides with the user i; T_j is the user frequency and T the time base, of which the user periods $\{T_m\}$ are integer multiples. Δ_j indicates the time duration of the pulse and is restricted by $\Delta_j < T_j, \forall_{j \in M}$ and F_{ij} is the greatest common divisor of the integer numbers of the relative ith and jth time base and user periods related as $T_m = TN_m \mid m \in \{i, j\}$.

The transmission of the signal from receiver to the clinical back-end can be performed using any of the existing wireless communication channels like orthogonal frequency division and multiplexing (OFDM), spread spectrum (SS), Time/frequency/code division multiplexing and accessing (T/F/C DMA) and UWB transmissions [35].

Having delineated the architectural overview, the experimental test bed and its results are analyzed in the following section.

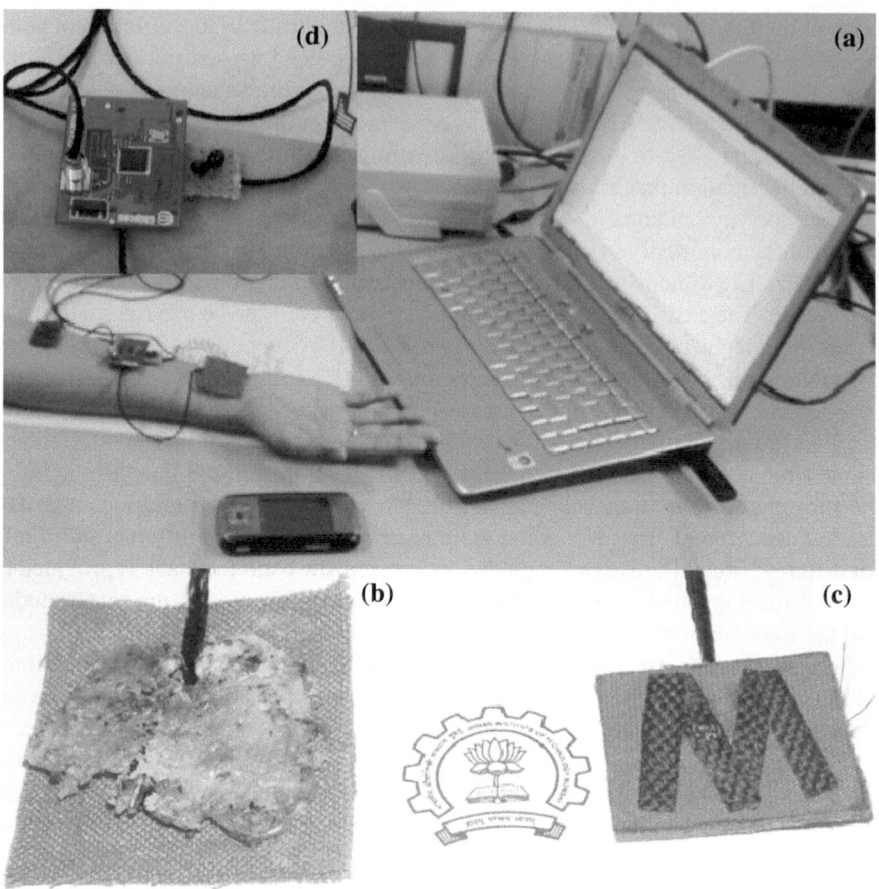

Fig. 6.5 a Experimental test bed for the pervasive computing WBAN architecture showing measurement of the radial artery using **b** tuPOY sensor and transmitting it with **c** tuPOY composite antenna. **d** The preprocessing of the signal after capturing by sensors for transmission to the antenna is performed in the OBTC

6.1.7 Experimental Test Bed

The WBAN architecture is validated by experimental trials as illustrated in Fig. 6.5. Figure 6.5a consists of the proposed tuPOY sensor for pulse analysis, an on-body transmission circuit, an on-body tuPOY composite antenna, a wireless transmission device and a back-end peripheral pulse analyzer.

The tuPOY sensor and the MPTA tuPOY antenna[2] are shown in Fig. 6.5b, c, respectively along with the OBTC in Fig. 6.5d. Due to the operational limitations at

---

[2]Developed at the SPANN Laboratory, Department of Electrical Engineering, Indian Institute of Technology, Bombay, India.

Table 6.4 Epidemiology of the extracted waveforms

Age (yrs)	I	II	III	IV	V	VI	VII	VIII
<20	191/62	115/23	212/28	91/65	72/29	129/31	68/2	62/3
20–40	118/38	128/17	127/75	193/79	82/52	339/80	77/12	23/31
40–60	168/11	52/29	425/113	25/22	412/61	378/58	145/29	242/50
>60	115/6	315/29	391/113	45/22	87/61	96/65	393/54	312/48

Table 6.5 Gender wise distribution of normal and diseased subjects

	Control subjects	Cirhossis of liver	Bronchial asthma	TB	MI
Male (%)	13.8	10.2	7.8	4.8	11.6
Female (%)	14.6	9.5	11.3	6.3	10

lab scale the shape and color of the sensor is not very uniform, however, a thin and transparent sensor can be easily made while commercial production.

The calibration for the eight different patterns of pulse waveform using tuPOY sensors is shown in Table 6.2. Five thousand patterns from more than two thousand volunteers in the age range of 18–57 years are acquired as shown in Table 6.4. The Genderwise demographics of plethysmographic patterns is provided in Table 6.5. The corresponding waveforms are generated by OBTC in the UWB modulated frequency range of 7.5–9.5 GHz.

The UWB signals are sent via tuPOY fibers to the feed of the 'M' shaped antennas. The experimentally measured characteristics of the tuPOY 'M' shape antenna are similar to the simulated results shown in Fig. 5.3 and hence are not repeated here, except for a major change noticed in radiation pattern as shown in Fig. 6.6. This is in correlation with the operational flow of the antenna as discussed in Chap. 5, where the electromagnetic energy directed toward the body is absorbed by the ground plane of the antenna. In addition to the above advantage, the captured heat generated from the absorbed radiation, is transmitted to the power generation unit via tuPOY conductors, which assist in the power generation of the system.

The radiated UWB data is received at the on-body receiver via a mobile[3] phone. The device is tuned to send the signal further to the pulse analyzer through UWB transmission. Since the experiments were conducted on lab scale and the distance between the back-end receivers is less, UWB transmission is used. However for actual operating architectures, alternate transmission schemes like OFDM or CDMA would be preferable, which can be adapted without any changes to the described architecture. The transmitted signals are received at the back-end, which mainly consists of the APM. Based on their morphologies the pulse patterns are classified into eight different classes.

---

[3]IMATE PDAL running on windows mobile 6.0 operating system.

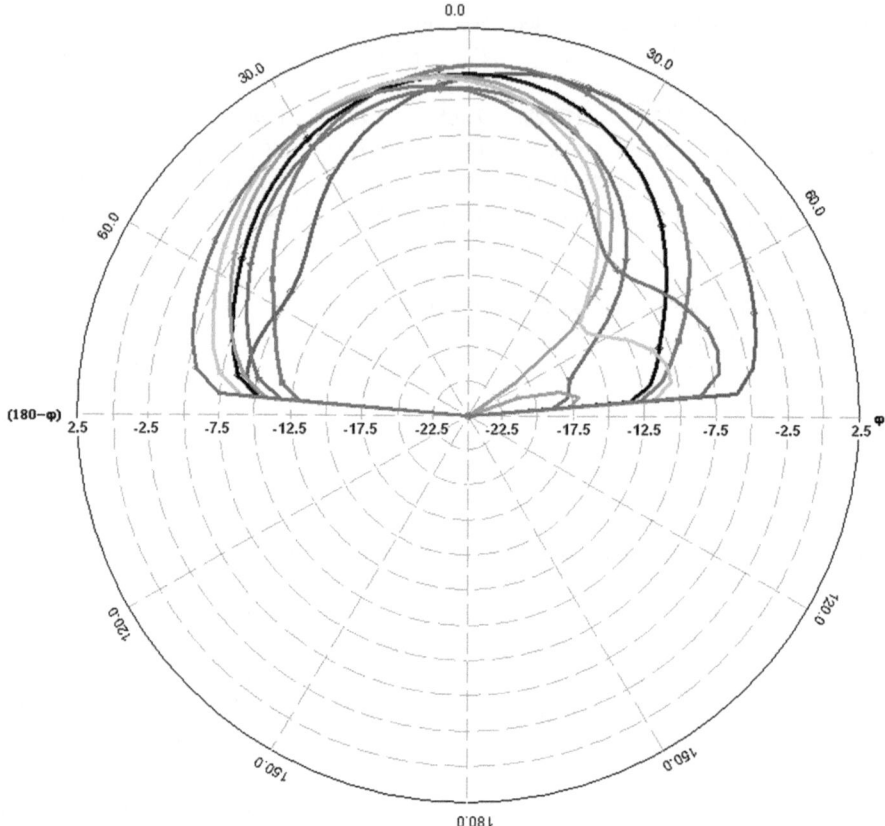

Fig. 6.6 Experimental results of tuPOY composite antenna measurement. The unidirectional radiation pattern confirms the absorption of harmful electromagnetic radiation directed towards the body

6.2 tuPOY: The Final Frontier

A metallic nonmetal, tuPOY charts a new dimension in engineering laying a pathway for alternative designs in processing applications. Juxtaposed with metallic properties, tuPOY delineates the spectrum of future processing. Researchers have embattled silicon as the beast of burden of the semiconductor family since 50 years. With advancement in nanotechnologies reaching its epitome, and scientists struggling to further tweak the size of MOS transistors, the need of the hour demands for a new material for processing applications. New processing elements in form of graphene, carbon nanotubes, and gallium nitrides, for the next generation computers have attracted heightened interest. Regarded once upon a time as 'killer' technologies and neo materials; they have yet to see the light of the day in any processing applications. Gallium nitride disappeared because of its self-heating properties while graphene and carbon nanotubes are believed to be heavily toxic. tuPOY as a metal retaining the robust properties of partially oriented yarn is potentially the 'silicon' of the future.

Partially oriented yarn being cheap, easily available, with known chemical properties and a well established production process forms a strong contender for a new processor evolution. The relatively easy accessibility of raw materials DMT and EG, the simplicity of polyester synthesis combined with the outstanding mechanical and chemical properties of POY, make it an ideal candidate for sensors and antennas. tuPOY is environmentally courteous due to its recyclability and its ability to be produced in closed loop with low toxic emissions. It is easily maintainable and has low specific weight with good strength, tensibility, and elasticity while being a synthetic fiber, resists variation due to stretch, shrinking, and wrinkles. tuPOY as a processing material has better abrasion resistance, drapability, and withstands bending stresses, unlike silicon and other alternatives.

The production mechanism of tuPOY enlightens the use of retardants and catalysts in a parallel domain for next generation development. The concept of thermal unstability is evolved and characterized by spectroscopic techniques, such as XRD, TEM, SEM, FTIR and NMR. Conformational structural information for understanding the morphology, the degree of crystallinity to confirm the textile properties of tuPOY, microscopic imaging to understand its metallic behavior and magnetic resonance in characterizing the ratio of the intermediates to determine the unstability are established with the support of these techniques in an unconventional demeanor. The innovative and judicious fabrication of the tuPOY as antennas and sensors offer satisfactory performance, reflecting almost negligible, harmful radiation, effects. A power generating unit designed with tuPOY at its core effectively generates power by harvesting energy from thermic domain and electromagnetic radiations, outlines a new dimension of operational power dynamics.

Paradigms presented from the perspective of pervasive computing are protruded in an innovative architecture as a case study WBAN. tuPOY is successfully tested for sensing, conducting and radiating applications for the radial pulse artery on more than 2000 volunteers. The case study presented here for the radial pulse analysis can be similarly implemented for any other waveforms of the human body and multitude of other un-envisaged applications.

Complimentary to organic polymers, tuPOY eliminates the need for actuators and significantly exceeds them in terms of stability and durability. Graphene and carbon nanotubes are deemed unsuited for use in logic and processing applications as their gate remains slightly conductive even when no voltage is applied to the base input. tuPOY overcomes this limitation by setting its own inherent threshold in form of microstress and this threshold moderation is used to leverage a switching action in tuPOY. Comprehensively equipped with these characteristics, tuPOY as the core element in the next generation processors is exploited in our future endeavors. The interchange phenomenon forms the mathematical backbone to quantify and correlate science of processing in tuPOY, establishing device parameters for transistor action.

Conclusively with the advent of tuPOY, an alternate dimension of a naturally occurring phenomenon since ages, is now scientifically and technically explorable.

References

1. Coyle, S., Lau, K.T., Moyna, N., O'Gorman, D., Diamond, D., Francesco, F.D., Costanzo, D., Salvo, P., Trivella, M.G., Rossi, D.D., Taccini, N., Paradiso, R., Porchet, J.A., Ridolfi, A., Luprano, J., Chuzel, C., Lanier, T., Revol-Cavalier, F., Schoumacker, S., Mourier, V., Chartier, I., Convert, R., De-Moncuit, H., Bini, C.: BIOTEX: biosensing textiles for personalized healthcare management. IEEE Trans. Inf. Technol. Biomed. **14**(2), 364–370 (2010)
2. Fort, A., Keshmiri, F., Crusats, G., Craeye, C., Oestges, C.: A body area propagation model derived from fundamental principles: analytical analysis and comparison with measurements. IEEE Trans. Antennas Propag. **58**(2), 503–514 (2010)
3. Chiti, F., Fantacci, R., Archetti, F., Messina, E., Toscani, D.: An integrated communications framework for context aware continuous monitoring with body sensor networks. IEEE J. Sel. A. Commun. 27, 379–386 (2009). http://dl.acm.org/citation.cfm?id=1649534.1649537
4. Baldus, H., Corroy, S., Fazzi, A., Klabunde, K., Schenk, T.: Human-centric connectivity enabled by body-coupled communications. IEEE Commun. Mag. **47**(6), 172–178 (2009)
5. Villalba, E., Salvi, D., Ottaviano, M., Peinado, I., Arredondo, M., Akay, A.: Wearable and mobile system to manage remotely heart failure. IEEE Trans. Inf. Technol. Biomed. **13**(6), 990–996 (2009)
6. Patel, S., Lorincz, K., Hughes, R., Huggins, N., Growdon, J., Standaert, D., Akay, M., Dy, J., Welsh, M., Bonato, P.: Monitoring motor fluctuations in patients with Parkinson's disease using wearable sensors. Trans. Inf. Techno. Biomed. 13(6), 864–873 (2009). http://dx.doi.org/10.1109/TITB.2009.2033471
7. Gatzka, C.D.: Pulse pressure where, how, and why? Am. J. Hypertens. 21(4), 376 (2008). http://www.biomedsearch.com/nih/Pulse-pressure-where-how-why/18369356.html
8. O'Rourke, M.F., Pauca, A., Jiang, X.J.: Pulse wave analysis. Br. J. Clin. Pharmacol. 51(6), 507–522 (2001). (Online). Available: http://dx.doi.org/10.1046/j.0306-5251.2001.01400.x
9. Khan, I., Hall, P.: Multiple antenna reception at 5.8 and 10 GHz for body-centric wireless communication channels. IEEE Trans. Antennas Propag. **57**(1), 248–255 (2009)
10. Farserotu, J., Hutter, A., Platbrood, F., Ayadi, J., Gerrits, J., Pollini, A.: UWB transmission and MIMO antenna systems for nomadic users and mobile pans. Wirel. Pers. Commun. 22, 297–317 (2002). http://dl.acm.org/citation.cfm?id=609293.609698
11. Minseok, K., Takada, J.: Statistical model for 4.5-GHz narrowband on-body propagation channel with specific actions. IEEE Antennas Wirel. Propag. Lett. **8**, 1250–1254 (2009)
12. Reusens, E., Joseph, W., Latre, B., Braem, B., Vermeeren, G., Tanghe, E., Martens, L., Moerman, I., Blondia, C.: Characterization of on-body communication channel and energy efficient topology design for wireless body area networks. IEEE Trans. Inf. Technol. Biomed. **13**(6), 933–945 (2009)
13. Sani, A., Alomainy, A., Palikaras, G., Nechayev, Y., Hao, Y., Parini, C., Hall, P.: Experimental characterization of UWB on-body radio channel in indoor environment considering different antennas. IEEE Trans. Antennas Propag. **58**(1), 238–241 (2010)
14. Tewari, K., Sundaram, K.: Digital computer simulation of pulse wave transmission in arteries. Med. Biol. Eng. Comput. 9, 297–304 (1971). http://dx.doi.org/10.1007/BF02474083
15. Jeon, Y.J., Kim, J.U., Lee, H.J., Lee, J., Ryu, H.H., Lee, Y.J., Kim, J.Y.: A clinical study of the pulse wave characteristics at the three pulse diagnosis positions of Chon, Gwan and Cheok. Evid.-Based Complement. Altern. Med. 2011, 9 (2011)
16. Yoon, Y.Z., Lee, M.H., Soh, K.S.: Pulse type classification by varying contact pressure. IEEE Eng. Med. Biol. Mag. **19**(6), 106–110 (2000)
17. Tartiere, J.M., Logeart, D., Beauvais, F., Chavelas, C., Kesri, L., Tabet, J.Y., Cohen-Solal, A.: Non-invasive radial pulse wave assessment for the evaluation of left ventricular systolic performance in heart failure. Eur. J. Heart Fail. 9(5), 477–483 (2007). http://eurjhf.oxfordjournals.org/content/9/5/477.abstract
18. Min, G., Rowe, D.: Conversion efficiency of thermoelectric combustion systems. IEEE Trans. Energy Convers. **22**(2), 528–534 (2007)

19. Özkan, G., Ürkmez, G., Özkan, G.: Application of Box-Wilson optimization technique to the partially oriented yarn properties. Polym. - Plast. Technol. Eng. 42(3), 459–470 (2003). Cited by (since 1996) 5. http://www.scopus.com/inward/record.url?eid=2-s2.0-0042155545&partnerID=40&md5=8d844cfbe181b6ebf201aaeda2803d42
20. Gerrits, J., Farserotu, J., Long, J.: Low-complexity ultra-wide-band communications. IEEE Trans. Circuits Syst. II: Express Br. 55(4), 329–333 (2008)
21. Kim, E.C., Park, S., Cha, J.S., Kim, J.Y.: Improved performance of UWB system for wireless body area networks. IEEE Trans. Consum. Electron. 56(3), 1373–1379 (2010)
22. Park, J., Park, S., Shin, D., Park, D.: Performance analysis of layered and blended organic light-emitting diodes. In: International Conference on Numerical Simulation of Optoelectronic Devices: NUSOD'08, September 2008, pp. 23–24 (2008)
23. Gross, R.A., Kalra, B.: Biodegradable polymers for the environment. Science 297(5582), 803–807 (2002). http://www.sciencemag.org/content/297/5582/803.abstract
24. Sanz Izquierdo, B., Miller, J.A., Batchelor, J.C., Sobhy, M.I.: Dual-band wearable metallic button antennas and transmission in body area networks. IET Microw. Antennas Propag. 4(2), 182–190 (2010)
25. Henniker, J.: Triboelectricity in polymers. Nature 196, 474 (1962)
26. Swanson, B.W., Somers, E.V., Heikes, R.R.: Optimization of a sandwiched thermoelectric device. J. Heat Transf. 83(1), 77–82 (1961). http://link.aip.org/link/?JHR/83/77/1
27. Sherman, B., Heikes, R.R., Ure, J.R.W.: Calculation of efficiency of thermoelectric devices. J. Appl. Phys. 31(1), 1–16 (1960). http://link.aip.org/link/?JAP/31/1/1
28. Snyder, G.J., Ursell, T.S.: Thermoelectric efficiency and compatibility. Phys. Rev. Lett. 91, 148301 (2003). http://link.aps.org/doi/10.1103/PhysRevLett.91.148301
29. Declercq, F., Rogier, H.: Active integrated wearable textile antenna with optimized noise characteristics. IEEE Trans. Antennas Propag. 58(9), 3050–3054 (2010)
30. Klemm, M., Troester, G.: Textile UWB antennas for wireless body area networks. IEEE Trans. Antennas Propag. 54(11), 3192–3197 (2006)
31. Sanz Izquierdo, B., Batchelor, J.C., Sobhy, M.I.: Button antenna on textiles for wireless local area network on body applications. IET Microw. Antennas Propag. 4(11), 1980–1987 (2010)
32. Hernandez, M., Kohno, R.: Ultra low power UWB transceiver design for body area networks. In: 2nd International Symposium on Applied Sciences in Biomedical and Communication Technologies: ISABEL 2009, pp. 1–4 (2009)
33. Singh, S., Ziliotto, F., Madhow, U., Belding, E.M., Rodwell, M.J.W.: Blockage and directivity in 60 GHz wireless personal area networks: from cross-layer model to multihop MAC design. IEEE J. Sel. Areas Commun. 27(8), 1400–1413 (2009)
34. Li, M., Lou, W., Ren, K.: Data security and privacy in wireless body area networks. IEEE Wirel. Commun. 17(1), 51–58 (2010)
35. Otal, B., Alonso, L., Verikoukis, C.: Highly reliable energy-saving MAC for wireless body sensor networks in healthcare systems. IEEE J. Sel. A. Commun. 27, 553–565 (2009). http://dl.acm.org/citation.cfm?id=1649534.1649550

Index

A

Abrasion resistance, 1
Acidolysis, 14
Agglomerates, 7
Alcoholysis, 14
Antenna Efficiency, 47
Antenna substrates, 39
Aperture coupled patch antennas, 4
Aracon, 3
Aramid, 3
Aromatic polyamide, 58
Artificial intelligence core, 8
Attenuation coefficient, 7
Automated physician machine, 3

B

Bending vibrations, 26
Bis-β-hydroxyethyl terephthalate, 13
Bis-hydroxyl ethyl terephthalate, 13, 24
Bragg angle, 22
Briquettes, 15
Buckling, 6
Buckytubes, 8

C

Capacitive, 7
Carbo-hydroxyethoxy, 13
Carbomethoxy, 13
Carbon nanotubes, 6, 67
Carnot efficiency, 58
Characteristic impedance, 56
Chemical stability, 16
Chemically bonded, 24
Chirality, 6
Circularly polarized, 5

Composite textile antenna, 8
Conducting fibers, 39
Conjugated Matching Efficiency, 47
Conjugated polymers, 4
Constantum, 58
Copper grid, 21
Covalent bond, 26
Crystallinity, 19

D

Dc conductivity, 7
Denier value, 16
Diethylene Glycol, 24
Diffractions, 8
Diffusivity, 6
Dimethyl Terephthalate, 13
Dipole moment, 41
Drapability, 1, 5
Dual frequency modulation, 57
DuPont, 3
Duroid, 4

E

Electrical ambiance, 8
Electrical resonance, 27
Electromagnetic radiation, 61
Electron delocalization, 5
Electron microscopy, 19
Electrostriction, 5
Energy barrier, 2, 13, 16
Energy gap, 6
Esterification, 13
Ethylene Glycol, 13
Exchange reaction, 14
Extensible, 3

© Springer India 2016
H.D. Mustafa et al., *tuPOY: Thermally Unstable Partially Oriented Yarns*,
Advanced Structured Materials 23, DOI 10.1007/978-81-322-2632-1

Q
Quarter wavelength, 4

R
Radial artery, 2, 52, 55
Radial pulse signal, 61
Radiation Efficiency, 47
Radio waves, 24
Raw silk, 1, 40
Rayon, 1
Redox behavior, 5
Residual inductance, 3
Residual microstress, 36, 55
Resin, 8
Retardant, 13

S
Satin 5, 4
Schrodinger, 29
Secondary electrons, 21
Seebeck effect, 58
Sensor arrays, 5
Sensor patch volume, 36
Shieldit, 4
Single-walled, 6
Smart materials, 9
Songket, 4
Specific absorption rate, 60
Spectral fingerprint, 25
Spectrometric, 2
Spectroscopy, 19
Spread spectrum, 63
Static electricity, 1
Stretching vibrations, 26
Synthetic textiles, 53

T
Test bed, 63

Textile antennas, 39
Thermal conductivity, 58
Thermal energy, 8, 42
Thermal unstability, 2
Thermocouples, 58
Thermodynamics, 8, 30
Thermoelectric generators, 58
Thermopiles, 58
Transesterifications, 13
Transition probabilities, 32
Transmission Electron Microscope, 20
Transmittance rate, 42
Triethylene Glycol, 24
TuPOY, 8

U
Ubiquitous on-body computing, 51
Unreacted Functional Units, 31
Unstable state, 9

V
Van Der Waals forces, 6
Vibration frequency, 41
Vibrational energy, 26
Vibrational resonance, 26
Vinylidene fluoride, 5
Voltage Source Efficiency, 47

W
Wearable technology, 56
Wireless body area network, 2, 52
Wireless communications, 51

X
X-ray Diffraction, 19